Anatomy of a Giant Carbonate Reservoir: Fullerton Clear Fork (Lower Permian) Field, Permian Basin, Texas

AAPG Studies in Geology 63
Report of Investigations Series No. *276*

Edited By
Stephen C. Ruppel

Co-published by
The American Association of Petroleum Geologists
and
Bureau of Economic Geology

ISBN13: 978-0-89181-069-8

AAPG Editor: Stephen E. Laubach
AAPG Geoscience Director: James B. Blankenship
BEG Director: Scott W. Tinker

COVER: Fence diagram showing porosity variations defined from 3-D modeling of the lower Clear Fork, Wichita, and Abo units across the Fullerton Clear Fork Unit field area. Lowest porosity shown in blue; highest in red.

This publication is available from:

The AAPG Bookstore
P.O. Box 979
Tulsa, OK U.S.A. 74101-0979
Phone: 1-918-584-2555 or 1-800-364-AAPG (U.S.A. only)
E-mail: bookstore@aapg.org
www.aapg.org

U.T. Bureau of Economic Geology Bookstore
University Station, Box X
Austin, TX 78713-8924, U.S.A.
Phone: 512.471.7144
Toll-free phone: 1.888.839.4365 (USA)
Toll-free fax: 1.888.839.6277 (USA)
E-mail: pubsales@beg.utexas.edu
http://www.beg.utexas.edu/pubs/pubsales.php

Geological Society Publishing House
Unit 7, Brassmill Enterprise Centre
Brassmill Lane, Bath BA13JN
United Kingdom
Phone: +44-1225-445046
Fax: +44-1225-442836
E-mail: sales@geolsoc.org.uk
www.geolsoc.org.uk

The American Association of Petroleum Geologists Books Refereeing Procedures

The Association makes every effort to ensure that the scientific and production quality of its books matches that of its journals. Since 1937, all book proposals have been refereed by specialist reviewers as well as by the Association's Publications Committee. If the referees identify weaknesses in the proposal, these must be addressed before the proposal is accepted.

Once the book is accepted, the Association Book Editors ensure that the volume editors follow strict guidelines on refereeing and quality control. We insist that individual papers can only be accepted after satisfactory review by two independent referees. The questions on the review forms are similar to those for the AAPG Bulletin. The referees' forms and comments must be available to the Association's Book Editors on request.

Although many of the books result from meetings, the editors are expected to commission papers that were not presented at the meeting to ensure that the book provides a balanced coverage of the subject. Being accepted for presentation at the meeting does not guarantee inclusion in the book.

More information about submitting a proposal and producing a book for The American Association of Petroleum Geologists can be found on its web site: www.aapg.org.

About the Editor

Stephen C. Ruppel is a Senior Research Scientist at the Bureau of Economic Geology in Austin where he specializes in the sedimentological and geochemical characterization of Paleozoic and Mesozoic carbonate and shale successions. He has published more than 200 scientific papers and reports on these topics. Prior to coming to the Bureau in 1981, he worked at McGill University in Montreal, Canada and Chevron Oil Co. in New Orleans. He holds the Ph.D. in Geology from the University of Tennessee at Knoxville as well as BS and MS degrees in Geology from the University of Illinois at Urbana and University of Florida at Gainesville, respectively. At the University of Texas, he currently teaches Advanced Reservoir Geology in the Jackson School of Geosciences for geology and engineering graduate students. For most of his career, his research has focused on shallow water carbonate reservoir systems. His interests in carbonate research center mostly around the application of sequence stratigraphy to defining facies architecture and the use of stable and radiogenic isotopes for dating of sedimentary rock successions. He recently directed the Bureau's industry-supported, Permian Basin Synthesis Program consortium. The research conducted during this program forms the basis for this book. Although he continues to conduct research on carbonate systems, his current research is directed toward defining key components of rock attribute variability in hydrocarbon-bearing shale successions. He is the program director of the Bureau's Mudrocks Systems Research Laboratory.

Table of Contents

Note to the Reader

This publication is a continuation of AAPG's efforts to combine the best of both digital and print publications for its members. Studies 63, Anatomy of a Giant Carbonate Reservoir: Fullerton Clear Fork (Lower Permian) Field, Permian Basin,Texas, consists of two sections.

The print portion of the volume consists of all six full-length papers. The digital portion of the volume contains these papers, and four appendices in pdf format as well as the appendices' images in their original file formats.

The dual digital and print format allows authors greater scope for both length of manuscripts and number of color figures, and provides the reader much more versatility in using the content. For example, figures can be used directly for making slides or Microsoft PowerPoint© presentations, and can be enlarged on the screen as needed for easier viewing. Individual papers can be easily printed for later reading, and key word searching through the papers is possible.

Finally, this format allows AAPG the flexibility to publish a volume that contains material that will appeal to a wider audience with additional interests.

Ruppel, S. C., 2012, Summary of reservoir characterization studies in the Fullerton Clear Fork reservoir, *in* S. C. Ruppel, ed., Anatomy of a giant carbonate reservoir: Fullerton Clear Fork (Lower Permian) field, Permian Basin, Texas: Studies in Geology 63, p. 1–4.

Introduction

Summary of Reservoir Characterization Studies in the Fullerton Clear Fork Reservoir

Stephen C. Ruppel
Bureau of Economic Geology, The University of Texas at Austin, Austin, Texas, U.S.A.

ABSTRACT

Despite declining production rates, existing reservoirs in the United States contain large quantities of remaining oil and gas that constitute an enormous target for improved diagnosis and imaging of reservoir properties. The resource target is especially large in carbonate reservoirs, where conventional data and methodologies are normally insufficient to resolve critical scales of reservoir heterogeneity. The objectives of the research described in this volume were to develop and test such methodologies for improved imaging, measurement, modeling, and prediction of reservoir properties in carbonate hydrocarbon reservoirs. The focus of the study is the Permian Fullerton Clear Fork reservoir of the Permian Basin of west Texas. This reservoir is an especially appropriate choice because the Permian Basin is the largest oil-bearing basin in the United States and, as a play, Clear Fork reservoirs have exhibited the lowest recovery efficiencies of all carbonate reservoirs in the Permian Basin.

OVERVIEW

The elements of the research conducted on Fullerton field are presented in six chapters. In the first chapter, Ruppel and Jones describe the depositional facies and sequence stratigraphy of the reservoir on the basis of integrated study of outcrops in analogous outcrops in west Texas and subsurface data from Fullerton field. They also describe the key steps in using outcrop- and core-based facies-stacking patterns and wireline-log-response relationships to develop a robust reservoir framework having sufficient resolution and accuracy to form the basis for reservoir modeling and simulation. In addition, they document the products of diagenesis and, using three-dimensional (3-D) relationships of geochemical and porosity-distribution data, propose models for dolomitization and porosity development. Finally, they describe the use of two modern techniques for improved reservoir imaging: borehole imaging logs for improved definition of facies distribution and cycles and 3-D seismic data for improved definition of the distribution of reservoir porosity.

Jones and Lucia document the nature and distribution of reservoir rock fabrics in the reservoir in the

DOI:10.1306/13341535St63698

second chapter. Knowledge of rock fabrics—groupings of rock types with similar petrophysical relationships—is critical for accurate modeling of permeability and saturation. Jones and Lucia show that each of the three major stratigraphic units in the reservoir contains distinct rock-fabric types that belong to three petrophysical classes. These classes vary substantially in petrophysical character. For example, rocks from petrophysical class 3 exhibit two orders of magnitude lower permeability for a given porosity value than do rocks of petrophysical class 1. Rock-fabric types and petrophysical classes were mapped throughout the field, providing an improved basis for calculating the permeability and original oil saturation in the reservoir.

In the third chapter, Kane and Jennings describe the procedures and methods used to develop a robust calculation of well-log porosity for each well in the field. These include the processes used to assemble and quality check the original wireline-log data for creating a workable database and the methodology used to create a final calculated porosity. Because calculations of reservoir permeability and saturation are based on porosity, this process is fundamental for accurate reservoir characterization and modeling.

Lucia and Kane discuss the approaches used to calculate permeability from wireline logs. As the fourth chapter documents, accurate permeability calculation depends on two primary factors: porosity and rock fabric. As such, their work is built on the procedures documented by Kane and Jennings for porosity determination and interpretation and distribution of rock-fabric properties described by Jones and Lucia. Lucia and Kane also describe the approaches used to calculate initial water saturation using capillary pressure and rock-fabric data. This approach, which is commonly the best way to define original saturations in carbonates, is the key to calculating original oil in place and remaining oil resource.

In the fifth chapter, Zeng reports on research conducted to recondition 3-D seismic data and to develop an integrated wireline-log-3-D seismic-inversion model of porosity. The data reconditioning work demonstrates how relatively simple procedures can lead to greatly improved seismic resolution, in turn leading to a better definition of reservoir architecture and porosity distribution. Zeng also describes the procedures used to construct a robust wireline inversion model of porosity for the reservoir. The results of this work illustrate the tremendous untapped potential that exists in using 3-D seismic data for improved imaging of reservoir properties in incompletely exploited carbonate reservoirs.

Finally, Wang and Lucia present the methods and results of reservoir modeling and simulation activities undertaken during the project in the sixth chapter. The primary goal of these studies was to better define the distribution of original and remaining oil resources in the reservoir. A key part of this research was the construction of (1) a 2000-ac flow-unit model, and a flow-simulation sensitivity study of the model and (2) a 35,000-ac full-field static reservoir model. The simulation study produced an excellent image of waterflood performance and provides fundamental guidelines to predict the distribution of the remaining oil resources across the reservoir. The full-field model, which is based on the integration of research completed on stratigraphy, reservoir architecture, porosity distribution, rock fabrics, and petrophysics, provides an excellent modern image of the 3-D distribution of reservoir properties that will serve as a future guide for modeling effective recovery operations in the field. Calculations of original resource volume based on this new model will also provide a needed economic incentive for these plans.

Also included in the appendices of this volume are digital data files and illustrations (in .pdf and original file format on the CD in the back of this book) documenting and supporting the summaries just outlined. These data, which include core descriptions, stratigraphic cross sections, core photographs, thin-section photomicrographs, porosity and permeability calculations, conventional core analyses, and special core analyses, provide a wealth of fundamental geologic, petrophysical, and engineering data that are probably unequaled in published literature.

SUMMARY OF DATA AND FINDINGS

Data Collection and Interpretation

- quality-control procedures performed on 18,000 well logs
- core, log, and interpretive data assembled into an integrated database (Landmark OpenWorks) including:
  - 1500 well files (23,000 log files)
  - 30 core-analysis files
  - core-facies descriptions (27 wells)
  - stratigraphic tops (~45,000)
  - calculated porosity (733 wells)
  - calculated permeability (733 wells)
  - calculated saturation (733 wells)
  - flow-unit tops (265 wells)

Stratigraphic Architecture

- subsurface data integrated with outcrop models
- facies and cyclicity defined in 27 cored wells (14,383 ft [4384 m] of core)
- cycles and tops correlated in more than 850 wells (~45,000 tops)
- more than 1700 thin sections (19 wells) described
- core and log data integrated with 3-D and two-dimensional (2-D) seismic data
- sequence and cycle stratigraphic framework defined for entire field
- structure and thickness of major sequences mapped
- presence of karst features demonstrated
- model for porosity development developed
- 3-D seismic inversion used for improved imaging of porosity distribution

Petrophysical Modeling

- new conventional core analyses (705) obtained
- new special core analyses (30) obtained
- new thin sections for rock-fabric analysis (950) obtained
- core data analyzed from 63 cored wells
- log porosity calculated for 733 wells
- porosity-log-normalization procedures created and applied for entire field
- petrophysical classes defined and mapped throughout field
- porosity-permeability transforms defined for all petrophysical classes
- original saturation model constructed from capillary pressure and petrophysical class data

Interpretation of Three-Dimensional Seismic Data

- 3-D and 2-D data reconditioned for phase character and resolution
- 222 mi (357 km) of 2-D and 33,000 ac of 3-D data integrated with wireline and core data
- time structure and isochrons of major zones mapped
- major sequences in 3-D and 2-D data volumes correlated
- relationships between deep structure and reservoir quality defined
- amplitude extractions used for reservoir-quality mapping
- robust relationship demonstrated between impedance and porosity
- wireline-log-3-D seismic-inversion model constructed

Reservoir Modeling

- constructed 2000-ac flow model composed of 39 flow units, 85 wells, 265 layers, and 3.2 million cells (150 × 150 × 3 to 5 ft [46 × 46 × 1 to 2 m] cells)
- constructed 35,000-ac full-field reservoir model composed of 35 stratigraphic surfaces and/or cycles, 730 wells, 380 layers, and 42 million cells (150 × 150 × 10 to 15 ft [46 × 46 × 3 to 5 m] cells)

Flow Simulation

- history match and sensitivity study performed for 1600-ac flow model
- flow simulation completed in Eclipse software
- simulation results used to characterize sweep characteristics of field

Definition of Opportunities for Drilling and Recompletion

- 3-D amplitude analysis completed and recommendations provided on eight proposed infill locations
- production streamline analysis of 15 proposed infill wells completed and alternatives recommended
- reservoir models and simulation results provide invaluable guides to additional drilling and recompletion opportunities

Definition of Resource Volumetrics

- volumetrics calculated for entire reservoir using multiple petrophysical and stratigraphic techniques

MAJOR FINDINGS

Stratigraphic Architecture

- The 3-D seismic data can provide inaccurate images of reservoir architecture.
- Porosity development is a function of early diagenesis.
- Borehole-image logs rival cores in providing facies and cyclicity data.
- Cycle-scale architecture is not definable in peritidal-facies tracts.
- Karst features are widespread but discontinuous and of uncertain impact on reservoir performance.
- Porosity development is partly controlled by deep structure.

- Permeability is a function of facies, but porosity is a function of diagenesis.

Petrophysics

- Peritidal successions (Wichita Formation) contain high-porosity, low-permeability rock fabrics.
- Subtidal successions (Lower Clear Fork and Abo formations) contain higher permeability rock fabrics than peritidal successions.
- Peritidal limestones are flow baffles; subtidal limestones are high-flow zones.
- Petrophysical classes and rock fabrics can be mapped throughout the field using a stratigraphic framework.
- A single porosity cutoff or permeability transform is inadequate.
- Permeability models must consider rock-fabric distribution.

Reservoir Simulation

- Simulation provides key insights to sweep and remaining oil distribution for much of the reservoir.
- Dominant issues controlling sweep efficiency are rock fabric, continuity, and completion coverage.
- Areas of poor history match and high water production are likely associated with zones of karsting in the Wichita Formation.

Three-Dimensional Geophysics

Porosity Characterization

- A strong relationship exists between amplitude and porosity.
- Progressive 3-D inversion provides excellent constraints on interwell and extrawell porosity distributions.
- Simple amplitude extractions provide a robust qualitative guide to interwell and extrawell porosity distributions.

Architecture

- Seismic response is controlled by porosity (facies and diagenesis).
- Seismic architecture must be vetted by geologic models.
- Definite evidence of continuing and differential fault motion through the Permian exists.

CONCLUSIONS

Collectively, the research described in this volume represents one of the most comprehensive studies of a major carbonate reservoir ever published. As such, it contains lessons of value for geoscientists, engineers, and managers, not only about the Fullerton Clear Fork reservoir and Permian reservoirs, but also about all carbonate platform reservoirs in the Unites States.

ACKNOWLEDGMENTS

The research documented here was co-funded by the U.S. Department of Energy (DOE), The University of Texas System, ExxonMobil Corporation, and the Reservoir Characterization Research Laboratory of the Bureau of Economic Geology. James Jennings, Ursula Hammes, James Gibeault, Florence Bonnaffé, Jean-Philippe Nicot, and Julia Gale provided thoughtful critical reviews of the manuscripts. I thank John Stout, Jeff Simmons, and Craig Kemp of Occidental Permian for providing important data for the study and for providing thoughtful insights into project planning. Steve Krohn and Amy Powell helped develop and promote the project within ExxonMobil. Terry Anthony and David Smith, also with ExxonMobil, provided ongoing operator liaison, data exchange, and input on project activities and data interpretation. Tim Hunt from The University of Texas System, West Texas Operations Office, helped with technology transfer activities and gathered new global positioning system location data for wells in the field. Dan Ferguson served as a contract officer for DOE. The support of this research by the Landmark Graphics Corporation via the Landmark University Grant Program is acknowledged. The publication was approved by the director of the Bureau of Economic Geology.

1

Ruppel, S. C., and R. R. Harrington, 2012, Facies and sequence stratigraphy: Critical tools for reservoir framework definition, Fullerton Clear Fork Reservoir, Texas, *in* S. C. Ruppel, ed., Anatomy of a giant carbonate reservoir: Fullerton Clear Fork (Lower Permian) field, Permian Basin, Texas: Studies in Geology 63, p. 5–48.

Facies and Sequence Stratigraphy: Critical Tools for Reservoir Framework Definition, Fullerton Clear Fork Reservoir, Texas

Stephen C. Ruppel
Bureau of Economic Geology, The University of Texas at Austin, Austin, Texas, U.S.A.

Rebecca R. Harrington
ExxonMobil Exploration Company, Houston, Texas, U.S.A.

ABSTRACT

Clear Fork reservoirs in the Permian Basin typically display a wide range of geologic and petrophysical properties that make the efficient recovery of hydrocarbons difficult. A key step in improving recovery efficiency is defining the patterns of variability in these rocks. The critical elements of variability that must be defined are facies, groupings of rocklike properties; and sequence architecture, the framework of facies variability. As in all carbonate reservoirs, rock-based studies must form a fundamental basis for characterizing and modeling facies and sequence architecture variability through the reservoir. Combined with wireline-log data, they provide a basis for defining both rock attribute distributions and reservoir framework.

At Fullerton field, we used 29 cores (>14,000 ft [>4270 m]), well logs from approximately 800 wells, three-dimensional seismic data, and outcrop data to define facies (rock attributes) and sequence stratigraphy (reservoir framework). The Fullerton reservoir section averages 500 ft (152 m) that can be subdivided into three stratigraphic units (Abo, Wichita, and Lower Clear Fork) and parts of two composite and six high-frequency sequences. At the base of the reservoir section, Abo rocks (sequences L1.1 and L1.2) consist of clinoformal, outer-platform, subtidal, fusulinid-crinoid packstones that exhibit locally excellent porosity and permeability characteristics but are highly variable in continuity. Wichita rocks were deposited in peritidal tidal-flat settings and consist of mud-rich facies that generally display poor continuity and commonly very low permeability and oil

DOI:10.1306/13341536St63698

saturation despite locally high porosity. Wichita rocks (sequences L1.2 and L2.0) are updip inner-platform equivalents of both partly underlying Abo and overlying Lower Clear Fork facies. Lower Clear Fork rocks (sequences L2.1 and L2.2) are dominantly middle-platform subtidal, grain-rich ooid-peloid packstone and grainstone facies that exhibit the best permeability and oil saturation properties.

Although basic facies distributions are defined by high-frequency sequence architecture, the reservoir framework must be based on the correlation of higher resolution depositional cycles. Because gamma-ray logs showed little or no relationship to facies and cyclicity, we calibrated porosity logs to cyclicity and used them to define 10 to 15 ft (3 to 5 m) cycles throughout the reservoir.

INTRODUCTION

The Clear Fork Group (Figure 1) in the Permian Basin comprises a thick (as much as 2500 ft [762 m]) succession of dominantly shallow-water-platform carbonates that were deposited across west Texas and New Mexico during the Early Permian (Leonardian). The reservoirs developed in these carbonates (Figure 2) have accounted for more than 3.2 billion bbl of oil production (Dutton et al., 2004)—more than 10% of the total recovered from the Permian Basin to date. As a whole, Leonardian reservoirs are estimated to have contained more than 14.5 billion bbl of oil at discovery, for a recovery efficiency of only about 22%, considerably less than the 32% average for carbonate reservoirs in the Permian Basin (Tyler and Banta, 1989; Holtz et al., 1992).

To recover the remaining oil in Leonardian reservoirs, operators must turn to increasingly sophisticated recovery technologies, for example, waterflooding, gas injection, and horizontal wells. To deploy these technologies effectively, however, it is critical that an accurate reservoir framework first be constructed to form the basis for modeling and interpreting past, present, and future recovery operations. This chapter describes the approaches used to develop such a framework in Fullerton Clear Fork field in west Texas. Like many mature Clear Fork Group reservoirs in the Permian Basin, Fullerton field has undergone a major decline in oil production rates and a major increase in water production rates for several years. The current total production from the field stands at about 310 million bbl, representing only about 25% of the calculated original oil in place. An understanding of the types of depositional and diagenetic facies, cycle and sequence stratigraphy, and architecture of these stratigraphic elements is the first crucial step to determine the probable oil distribution at discovery and define the best strategies for recovery of the sizable remaining oil volume.

METHODS

The focus of this study of the Fullerton reservoir is the Fullerton Clear Fork Unit, a unitized production area in Fullerton field operated by ExxonMobil Corporation. Data and interpretations presented in this chapter were derived from the investigation of 29 cores totaling 14,383 ft (4384 m), examination of more than 1700 rock thin sections, and correlation of about 45,000 stratigraphic tops in nearly 800 wells. The basic procedures used in this study to develop a full-field stratigraphic model and reservoir framework were as follows: (1) identify depositional facies and vertical facies-stacking patterns and cycles in cores, (2) calibrate facies and cycle patterns to wireline logs, (3) construct two-dimensional (2-D) cross sections of core-log data sections, (4) define 2-D cycle and sequence stratigraphy from core-log sections, and (5) extrapolate 2-D cycle and sequence correlations into three-dimensional (3-D) space by correlating all available well logs. Stratigraphic sequences were tied into available 3-D seismic data to check their accuracy and geometry. Conventional core analysis data were available for all cores; these data were used to determine relationships between facies, cyclicity, porosity, and permeability development. Stable isotope data were collected during the reservoir study and reported in Ruppel and Jones (2006). Additional data were collected for this report. Procedures are documented in Ruppel and Jones (2006).

Despite the emphasis placed on sequence- and cycle-stratigraphic concepts in establishing the reservoir framework, note that an understanding of rock properties is as important as reservoir architecture in building a reservoir model. Also note that, essentially by definition, sequences do not equate to facies. Throughout this chapter, reference is made to both time (sequence-stratigraphic) and rock (formation and facies) subdivisions of the succession to emphasize this point.

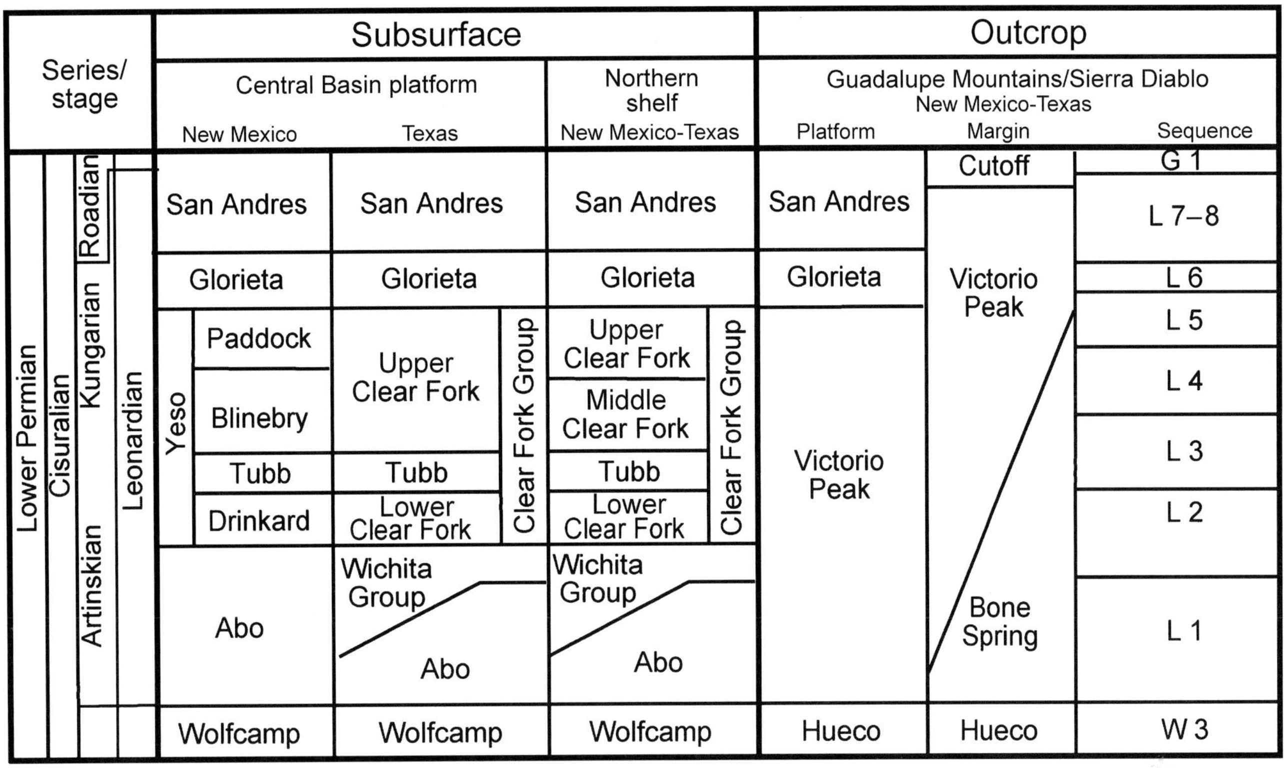

FIGURE 1. Chart showing the Leonardian stratigraphic section in the Permian Basin, including Clear Fork Group and analogous units in New Mexico and in outcrop. Modified from Ruppel (2002).

PREVIOUS WORK

Although detailed studies of Clear Fork reservoir successions are few, some important aspects of these rocks have been previously described. Mazzullo (1982) and Mazzullo and Reid (1989) presented overviews of lower Leonardian stratigraphy and depositional systems in the Midland Basin mostly based on studies of wireline logs and well cuttings. Presley and McGillis (1982) (see also Presley, 1987) documented the highly cyclic, predominantly evaporitic facies of the upper Leonardian Glorieta and Upper Clear Fork units in the innermost platform areas of the Texas Panhandle. Ruppel (1992, 2002) described the facies, cyclicity, and diagenesis in the Glorieta and Upper Clear Fork at Monahans Clear Fork field on the Central Basin platform and postulated that reservoir development was caused by cyclic deposition and diagenesis driven by episodic sea level rise and fall. Atchley et al. (1999) described Clear Fork facies at Robertson field at the northern end of the Central Basin platform and proposed a similar model for structural control over facies deposition and reservoir development. Ruppel et al. (2000) described outcrop equivalents of the producing subsurface Clear Fork reservoirs from outcrops in the Sierra Diablo Mountains of west Texas. Kerans et al. (2000) documented the depositional setting, facies, and stratal architecture of the Abo from outcrops in the Sierra Diablo (Figure 2) and showed that karsting has had a major effect on both the Abo and the overlying Lower Clear Fork succession. Ruppel and Jones (2006) presented an earlier analysis of Fullerton reservoir geology upon which this chapter is based.

GEOLOGIC SETTING

The Fullerton Clear Fork field is the largest of a large number of fields developed in the Leonardian series on the Central Basin platform of the Permian Basin (Figure 2). In some parts of the basin, the Leonardian is productive from all of its component stratigraphic units (Figure 1). The productive reservoir section at Fullerton field, however, is essentially restricted to the Lower Clear Fork, Wichita, and Abo stratigraphic units (Figure 3). By far, the bulk of the oil production has come from the lower two thirds of the Lower Clear Fork and the Wichita sections of the reservoir. Even where oil saturated, the Abo has proven difficult to exploit, partly because of its active water drive (as opposed to the pressure depletion drive that characterizes the overlying parts of the reservoir) and because

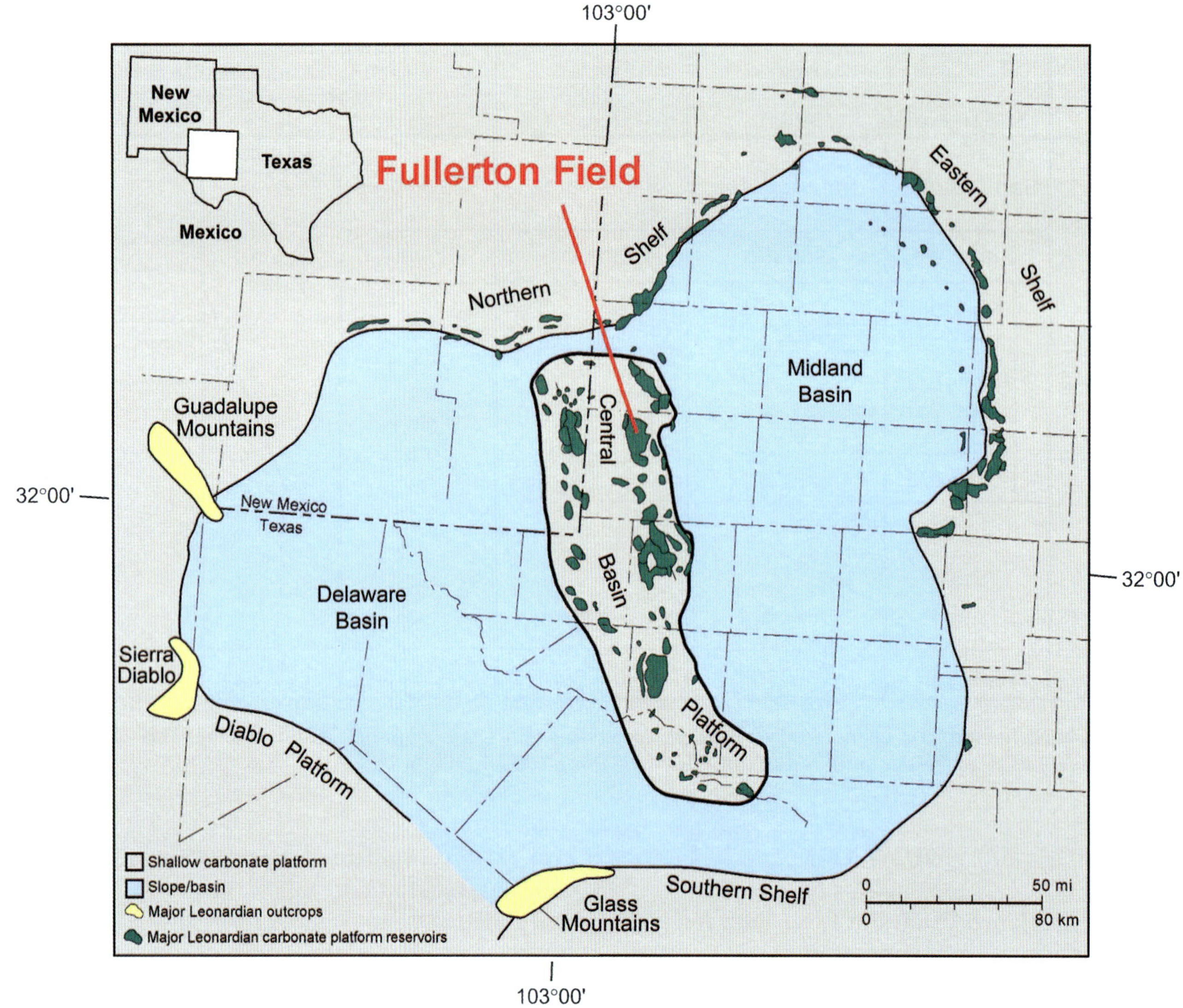

FIGURE 2. Regional map of the Permian Basin showing the location of Fullerton field and analogous outcrops in Sierra Diablo Mountains. Modified from Ruppel and Jones (2006).

it is at or near the oil-water contact throughout most of the field.

Regionally, the Leonardian succession is dominated by shallow-water platform carbonates. Each of the component Leonardian stratigraphic units contains updip peritidal tidal-flat carbonates and downdip subtidal carbonates. Mineralogy in each is dominated by dolomite and anhydrite. Calcite, in the form of limestone, is relatively uncommon; however, where present, it is most common in the lower part of the Leonardian and in the distal downdip sections (Ruppel, 2002). The reservoir section at Fullerton is generally consistent with this regional pattern, but it does contain a higher volume of limestone than most other Leonardian platform reservoir successions in the basin.

The Fullerton field is developed over a large compound structural high (Figure 4) that reflects the reactivation of deep-seated faults and differential uplift of the area that began in the Late Mississippian and continued into the Permian (Kluth and Coney, 1981). Evidence exists that, locally, these tectonic features continued to influence depositional patterns during the Leonardian (Jones and Ruppel, 2004). The Clear Fork reservoir seal is provided by evaporite-rich carbonates in the Lower Clear Fork, Tubb, Upper Clear Fork, and Glorieta.

The reservoir is currently drilled to well spacings of 40 to 10 ac and is under active waterflood. The greatest well density is in the northern part of the field (mostly 20- and 10-ac well spacings); poorer well control exists in the southern half of the field. In general, these

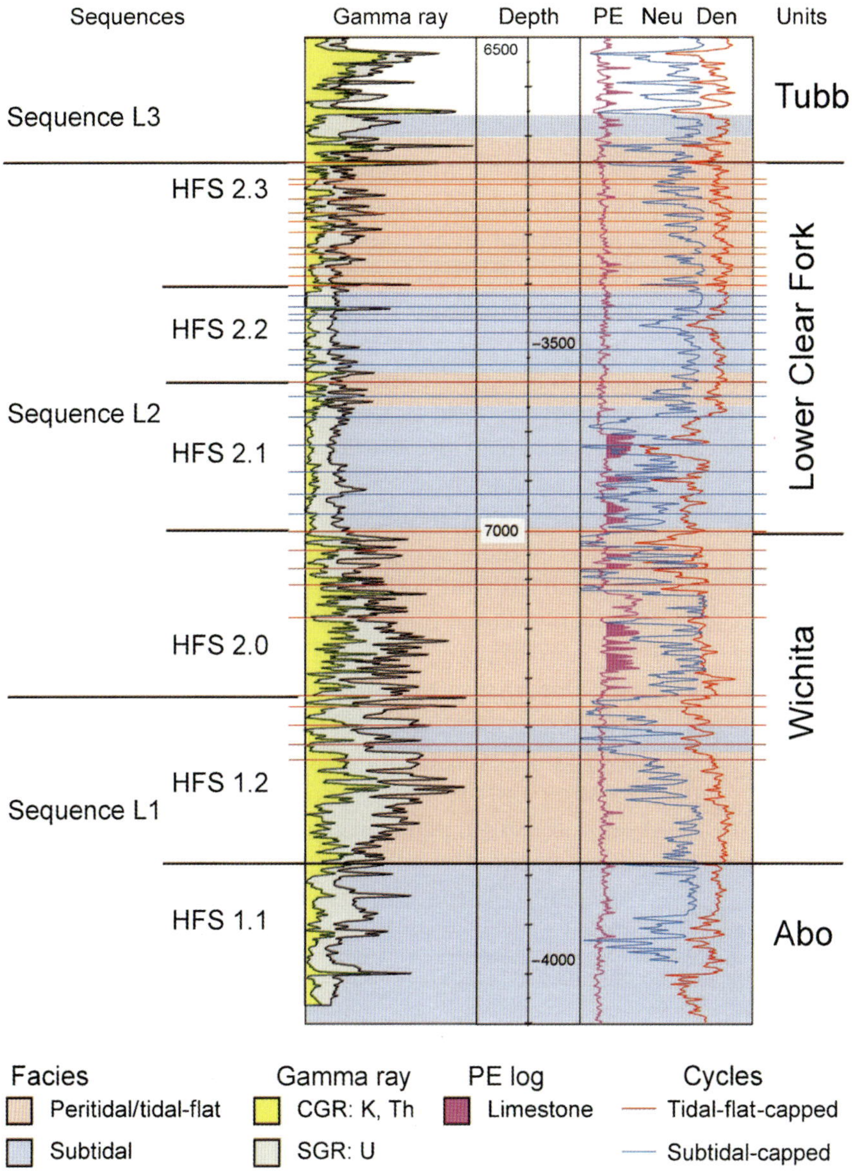

FIGURE 3. Type log of the producing reservoir section at Fullerton Clear Fork field showing cyclicity and general facies patterns. The shaded areas of PE (photoelectric) log indicate the local presence of limestone in dolomite-dominated sections. Zones of elevated corrected gamma-ray (CGR)-log response indicate the presence of silt and clay. Neu = neutron porosity log; Den = density porosity log; HFS = high-frequency sequence; SGR = spectral gamma ray. Modified from Ruppel and Jones (2006).

closely spaced wells provide good control for the definition of stratigraphic zones and reservoir attributes. In many parts of the field, especially along the western edge and in the southern half of the field, wells are represented only by poor-quality (old gamma-ray-neutron) wireline logs and cannot be correlated or interpreted with any precision.

OUTCROP FACIES, CYCLICITY, AND SEQUENCE ARCHITECTURE

The outcrops of Leonardian age in the Sierra Diablo Mountains in Hudspeth and Culberson counties (Figure 2), Texas, are excellent analogs for the producing reservoir section at Fullerton field and thus offer important insights into geologic controls on reservoir development and reservoir architecture. Leonardian outcrops in west Texas were originally assigned to the Victorio Peak Formation (King, 1942, 1965). However, recent studies by Ruppel et al. (2000) have extended subsurface nomenclature to these outcrops. Sierra Diablo outcrops include both shallow-water platform carbonates of the Lower Clear Fork Group and Wichita units (Fitchen et al., 1995; Ruppel et al., 2000) and the karsted platform-marginal subtidal carbonates of the Abo (Fitchen et al., 1995; Kerans et al., 2000). The styles of depositional architecture and facies development displayed in outcrops are closely similar to those encountered in the subsurface, making these outcrops invaluable aids for interpreting and modeling facies distributional patterns and stratigraphic architecture in subsurface reservoir successions. The stratigraphic framework defined by Fitchen et al. (1995) is now widely recognized as

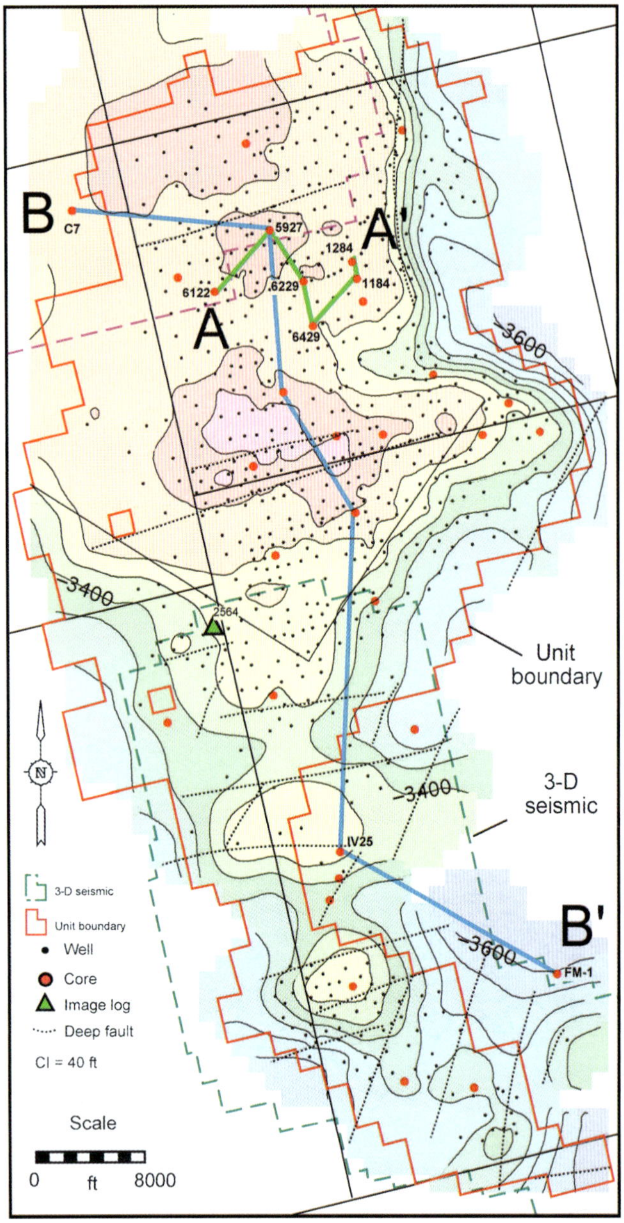

FIGURE 4. Structure of Fullerton Clear Fork field. Datum is a prominent subtidal flooding event near the base of the Tubb Formation (see Figures 3, 7). Cross section AA′ shown in Figure 7; cross section BB′ shown in Figure 15. CI = contour interval; 3-D = three-dimensional. Modified from Ruppel and Jones (2006).

definitive of the sequence stratigraphy of the Lower Leonardian succession (Figure 5). Similarly, high-resolution studies of Lower Leonardian platform facies cyclicity by Ruppel et al. (2000) have provided a basis for a better understanding of facies stacking and continuity in subsurface reservoir successions.

Abo Formation

Integrated studies of outcrops and subsurface data sets demonstrate that the Abo represents the basal depositional sequence (sequence L1) of the Leonardian (Fitchen et al., 1995). Studies of Abo outcrops in the Sierra Diablo (Fitchen et al., 1995; Kerans et al., 2000) defined three important aspects of this succession in the Permian Basin: (1) it consists of dominantly open-marine outer-platform facies, (2) it displays clinoformal architecture, and (3) it is overprinted by karst features (sinkholes, caves, cave fill, and collapse features) (Fitchen et al., 1995; Kerans et al., 2000). The dominance of clinoformal, outer-ramp, fusulinid-crinoid packstones and wackestones, and less common ramp-crest, ooid-peloid, grain-rich packstones and grainstones in the Abo contrasts with the flat-lying, alternating tidal-flat, and shallow subtidal wackestone-packstone successions of the Wichita and Lower Clear Fork. The top-lapping clinoforms of the Abo document rapid basinward progradation, a forced regression probably caused by a rapid fall in sea level. Karsting in the Abo, although initiated at the exposed top of the Abo during sea level fall, is manifested downsection in the Abo by caves and sinkholes, and upsection in the overlying Lower Clear Fork as collapse features. Where karsting and associated thickness variations are developed, mostly in platform-marginal settings, the contact between the Abo and Lower Clear Fork is a relatively sharp, undulating, and unconformable surface. Updip, karsting is less apparent, and the contact is less pronounced.

Lower Clear Fork Formation

Outcropping Lower Clear Fork and Wichita rocks in the Sierra Diablo represent a single depositional sequence (L2). Key elements of this succession are (1) basal backstepping tidal-flat deposits that locally fill relief on the underlying karsted Abo surface (top of sequence L1); (2) an updip L2 succession of amalgamated tidal-flat facies (Wichita equivalent); and (3) a cyclic downdip succession of alternating tidal-flat and mid-ramp subtidal facies (Lower Clear Fork equivalent).

Lower Clear Fork outcrops record gradual flooding of the previously exposed Abo platform and a continued, although punctuated, increase in accommodation and water depth. Downdip, Lower Clear Fork deposits are characterized by alternating peritidal tidal-flat deposits and subtidal skeletal wackestones and packstones that document cyclic rise and fall of sea level at the high-frequency sequence (HFS) and cycle scale. These cycles, which average about 20 ft (~6 m) in thickness, display consistent patterns of facies stacking (cycle-base skeletal wackestones and overlying peloidal grain-rich packstones) and appear to be

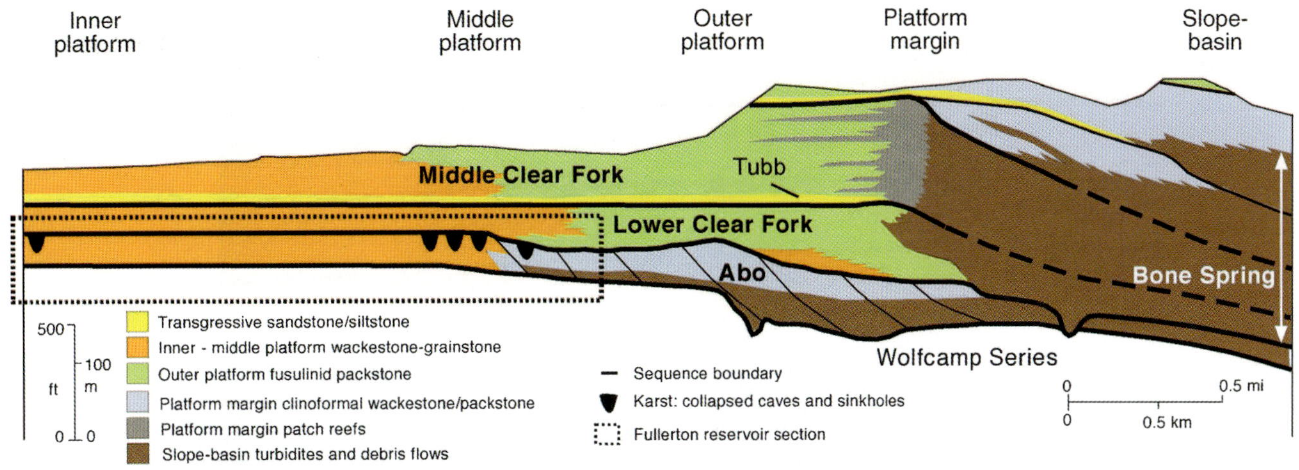

FIGURE 5. **Sequence stratigraphy of the Leonardian section in the Sierra Diablo Mountains. Based on Fitchen et al. (1995), Ruppel et al. (2000), and Kerans et al. (2000). Note relative position of Fullerton reservoir section.**

widely continuous (Ruppel et al., 2000). More recent studies show that these shallow subtidal platform rocks pass downdip into clinoformal fusulinid-crinoid wackestones and packstones of the outer ramp to slope less than 2 mi (<3 km) basinward. Updip, in the platform interior (landward), the Lower Clear Fork is characterized by amalgamated peritidal tidal-flat deposits. These updip tidal flats are analogous to the Wichita of the subsurface. The absence of shallow-water highstand deposits at the top of the Lower Clear Fork suggests a rapid, perhaps forced, regression, followed by exposure and possible erosion at the top of the L2 sequence.

Lower Clear Fork outcrops in the Sierra Diablo also provide important insights into the cycle architecture of analogous subsurface reservoir sections such as that at Fullerton (Ruppel et al., 2000). In tidal-flat-capped successions and, less commonly, in subtidal successions, 3 to 6 ft (1 to 2 m)-thick cycles are definable in vertical sections but, because of poor outcrop continuity, are not readily correlatable. However, subtidal cycle bundles, which average from 15 to 30 ft (5–9 m) in thickness, can be correlated at typical interwell distances. These bundles typically consist of upward-shallowing successions that have skeletal wackestones at their bases and peloid-ooid packstones at their tops.

FULLERTON RESERVOIR FACIES AND STRATIGRAPHY

The productive reservoir section at Fullerton field includes parts of the Abo, Wichita, and Lower Clear Fork formations (Figure 3). Both subsurface and outcrop studies demonstrate that the Abo and the lower part of the Wichita Formation are time-equivalent facies of the earliest composite (third-order) depositional sequence of the Leonardian (sequence L1). Sequence L2 comprises the upper part of the Wichita Formation (updip peritidal deposits) and the Lower Clear Fork Formation (downdip, dominantly subtidal deposits). The L3 sequence contains siliciclastics of the Tubb Formation and carbonates of part of the overlying Upper Clear Fork (Figure 3). These interrelationships among depositional sequences, facies, and stratigraphic units are depicted in Figure 6 for Fullerton field on the Central Basin platform.

Correlations of calibrated wireline-log suites from about 750 wells show that, across most of the area of the Fullerton Clear Fork Unit (Figure 4), stratigraphic units are relatively isopachous (Figure 7). This observation supports the interpretation that the depositional setting of the field area was a broad flat carbonate platform not unlike the modern-day Bahamas platform.

The same generally isopachous nature of the Lower Clear Fork Formation is apparent in 3-D and 2-D seismic data. These data, however, show that, by contrast, the Wichita and Abo vary considerably in thickness across the field. Wichita deposits thin and Abo deposits thicken to the east and southeast (Figure 8). Synthesis of core, wireline, and seismic data show that this reciprocal thickness relationship is the result of facies change: Wichita facies representing updip shallower water (inner platform) equivalent of the distal deeper water (outer platform to slope) Abo.

Facies and Depositional Setting

Facies encountered at Fullerton field are typical of those observed throughout most carbonate-platform

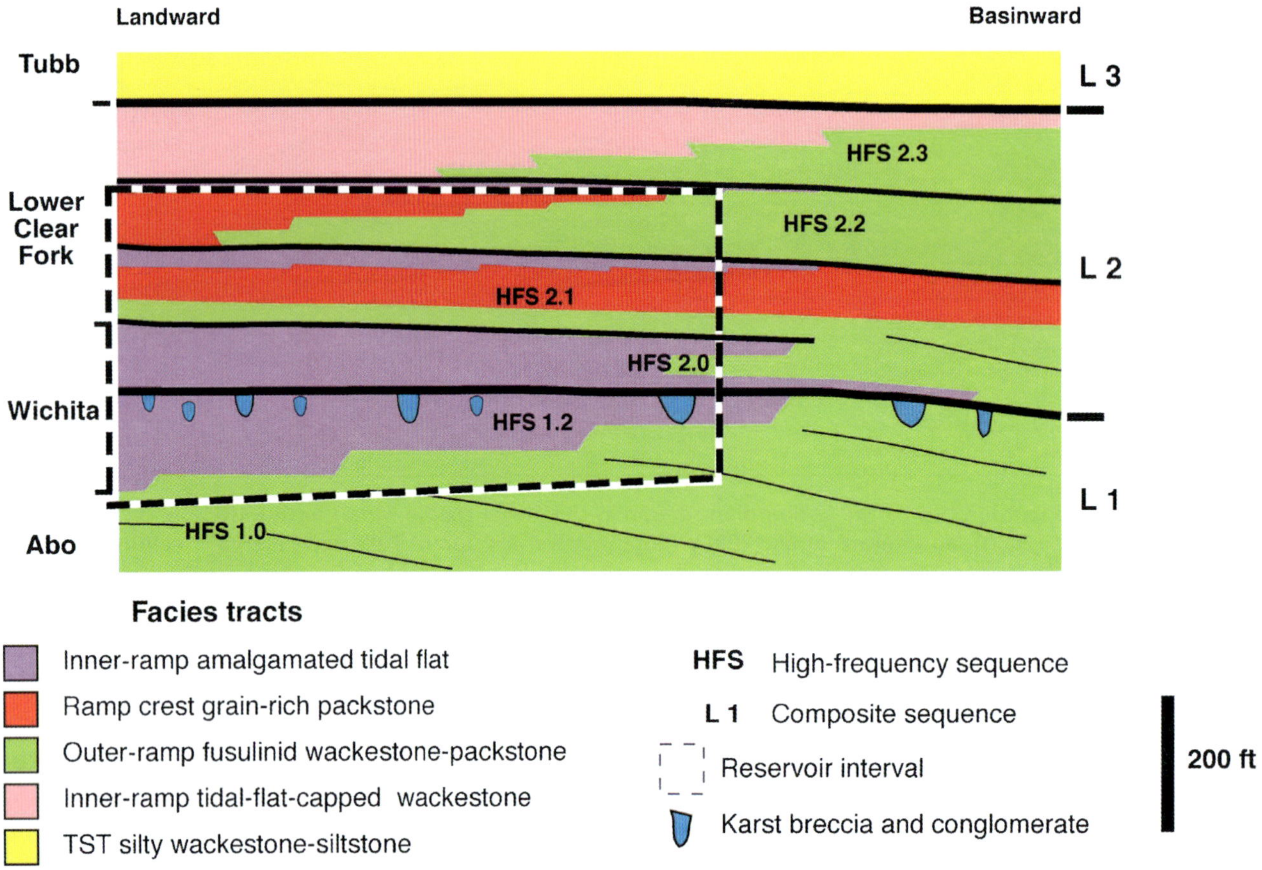

FIGURE 6. **Sequence-stratigraphic model of lower Leonardian succession at Fullerton field showing primary facies tracts and stratigraphic nomenclature. HFS = high-frequency sequence; TST= transgressive systems tract. Modified from Ruppel and Jones (2006).**

successions of Leonardian and early Guadalupian age in the Permian Basin (e.g., Bebout et al., 1987; Ruppel and Cander, 1988a, b; Garber and Harris, 1990; Longacre, 1990; Kerans et al., 1994; Kerans and Fitchen, 1995; Kerans et al., 2000; Kerans and Kempter, 2002; Ruppel and Bebout, 2001). Leonardian facies successions have been documented both in outcrop (Ruppel et al., 2000) and in the subsurface (Ruppel, 2002).

The facies defined in this study were defined with three goals in mind: (1) to document characteristics of distinct depositional settings, (2) to document widespread and potentially correlatable sediment packages, and (3) to create a sound geologic basis for assessing relationships between rock textures and fabrics and petrophysics. In many ways, it is the last of these goals that is most important because it provides the foundation for reservoir modeling. However, the facies characteristics used for subdivision are therefore dominantly small-scale matrix properties, for example, grain size, shape, and type. Larger scale features such as fractures, anhydrite nodules, and evidence of burrowing were also noted but not used for facies subdivision. Eleven depositional facies can be identified in the Fullerton reservoir succession (Table 1). Most are intergradational with one or more other facies and can be found in most parts of the reservoir succession.

Peritidal Mudstone-Wackestone Facies

Peritidal mudstone-wackestone facies are not only most abundant in the Wichita but are also locally common in the Lower Clear Fork. They are massive to parallel-laminated mud-rich rocks that only rarely contain any grains other than a few peloids (Figure 9A, B). They are most typically associated with exposed tidal-flat facies and clay-rich carbonate mudstone facies and are interpreted to represent peritidal deposition on a very low energy tidal flat. They are dominantly dolomitized and, where dolomitized, may display high porosities (as much as 15%). Because this porosity is caused by intercrystalline pores within fine-crystalline dolomite, however, permeability is generally low. Where limestone, they invariably exhibit very low porosity (typically <2%) and permeability. Gamma-ray signature is medium to high and highly variable

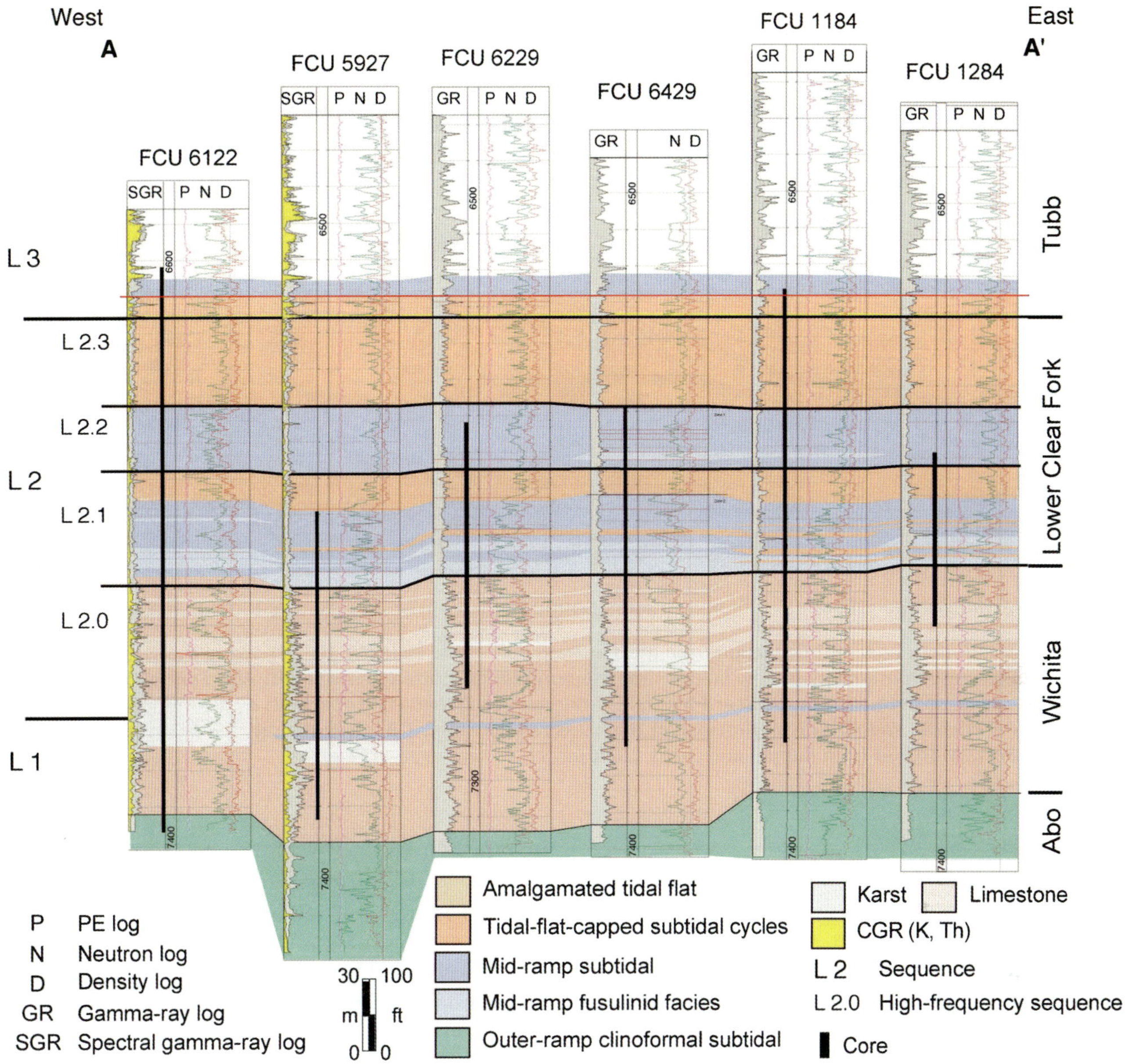

FIGURE 7. **Cross section (AA′) illustrating sequence stratigraphy and facies in the central part of Clear Fork reservoir at Fullerton field based on cores. The generally isopachous character of the reservoir section (Wichita–Lower Clear Fork) was consistent over most of the field area. Line of section shown in Figure 4. FCU = Fullerton Clear Fork Unit; CGR = corrected gamma ray.**

in these rocks because of the local presence of clay minerals.

Clay-Rich Carbonate Mudstone Facies

Clay-rich carbonate mudstone facies are found intimately associated with rocks of the peritidal mudstone-wackestone facies in the Wichita. They are typically dark gray to nearly black and 0.4 to 3 in. (1–8 cm) in thickness (Figure 9C). Where cyclic upward-shallowing successions can be identified, these deposits are found at or near cycle tops in both the Lower Clear Fork and the Wichita. Although they are contributors to the high gamma-ray response observed in the Wichita and in Lower Clear Fork peritidal intervals, because of their thinness, they cannot be discretely defined by logs and therefore are not correlatable. Their color, clay content and lack of apparent lateral continuity suggest that they were formed as local organic-rich ponds or stagnant pools on the tidal flat. They may act as local baffles to reservoir fluid flow but are probably very discontinuous laterally.

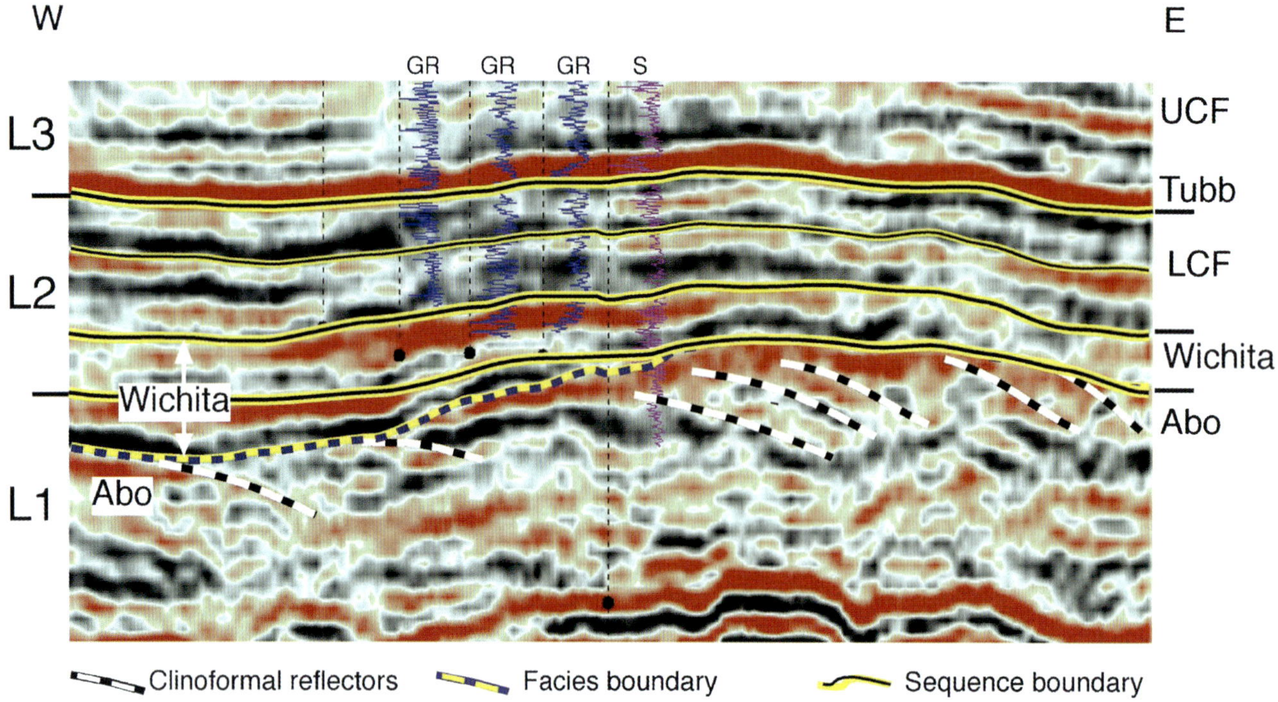

FIGURE 8. Three-dimensional seismic section from the southern part of the Fullerton field showing the seismic definition of the Clear Fork reservoir section. Yellow lines = timeline boundaries; dotted line = top of Abo Formation. Note that whereas the upper Wichita and Lower Clear Fork (LCF) intervals are essentially isopachous and continuous across the field, the lower Wichita and Abo display reciprocal thickness relationships. GR = gamma ray; S = acoustic log; UCF = Upper Clear Fork. Modified from Ruppel and Jones (2006).

Exposed Tidal-Flat Facies

Exposed tidal-flat facies show obvious evidence of exposure such as fenestral pores, pisolites, mudcracks, insect burrows, sheet cracks, tepee structures, and cyanobacterial or microbial laminations (Figure 9D, E). They are locally common in both the Lower Clear Fork and the Wichita. In the Lower Clear Fork, they define cycle tops; in the Wichita, their relative sea level significance is less apparent. Their sedimentary structures demonstrate, however, that they were formed during at least local sea level fall and exposure. Porosity is locally very high (as much as 30%) and is associated with fenestral pores and intercrystalline and interparticle pores. Permeability is lower than in grain-dominated subtidal rocks but can be significant where high porosities are present. Like the peritidal mudstone-wackestone facies, these rocks typically display somewhat elevated and variable gamma-ray response, which can sometimes be used to distinguish these rocks from overlying and underlying subtidal facies.

Peloid Wackestone Facies

The grain-poor peloid wackestone facies differ from rocks of the peritidal mudstone-wackestone facies in its association and sedimentary structures. Rocks are invariably associated with other demonstrably subtidal facies, are burrowed, and locally contain skeletal debris. These features suggest that they were deposited in low-energy subtidal settings. Dominant grains (commonly 80–150 μm in diameter) are probably fecal pellets created by infaunal burrowers. The absence of skeletal allochems supports a low-energy, perhaps restricted, setting.

Anhydrite nodules are locally common in these rocks, reflecting postdepositional entry of sulfate-bearing diagenetic fluids into permeability pathways created by burrowers. In many cases, anhydrite nodules occupy solution-widened vertical-burrow pathways that are surrounded by alteration halos (Figure 10A). These halos (which have been reported from most Permian platform-carbonate successions) commonly display a more grain-rich texture, higher porosity and permeability, and a depleted oxygen-isotope signature (Major et al., 1990; Ruppel and Bebout, 2001). At Fullerton, they are most commonly developed in HFS L2.2. This stratigraphic position (in the late highstand of the L2 composite sequence) is analogous in terms of accommodation to similar features documented from the Guadalupian Grayburg Formation at South Cowden field in Ector County, Texas (Ruppel and

Table 1. Characteristics of major depositional facies in the Fullerton Clear Fork Reservoir.

Facies	*Sedimentary Structures*	*Grain Types*	*Fossil Types*	*Mineralogy*	*Porosity and Pore Types*	*Depositional Setting*	*Cycle Position*
Peritidal mudstone-wackestone	Horizontal lamination, crinkly lamination, massive	Peloids	Rare mollusks	Dolomite*, calcite*, poikilotopic anhydrite	Low to high; intercrystalline	Tidal flat	Cycle top
Clay-rich carbonate mudstone	Horizontal lamination	None	Rare plant debris	Dolomite	Very low; intercrystalline	Tidal flat	Cycle boundary
Exposure-related mudstone-packstone	Fenestrae*, crinkly lamination*, insect burrows	Pisolites*, coated grains*, peloids, intraclasts	Rare mollusks	Dolomite*, poikilotopic anhydrite	Low to very high; fenestral, intercrystalline, interparticle	Tidal flat	Cycle top
Peloid wackestone	Burrow mottling*, burrows*	Peloids	Mollusks	Dolomite*, poikilotopic and nodular anhydrite	Low; intercrystalline, vuggy	Inner ramp, middle ramp	Variable
Peloid packstone	Burrow mottling*, burrows*	Peloids	Rare mollusks	Dolomite*, poikilotopic and nodular anhydrite	Low to moderate; intercrystalline, vuggy	Middle ramp, ramp crest, outer ramp	Cycle top
Ooid-peloid grain-dominated packstone-grainstone	Parallel lamination, inclined lamination, burrow mottling; good sorting	Ooids, peloids,		Dolomite*, poikilotopic anhydrite	Moderate to high; intergranular, vuggy (oomoldic)	Ramp crest*, outer ramp, middle ramp	Cycle top
Fusulinid wackestone-packstone	Burrow mottling	Peloids, pellets	Fusulinids; uncommon crinoids	Dolomite*, poikilotopic and nodular anhydrite	Low to moderate; intercrystalline, vuggy (oomoldic)	Outer ramp	Cycle base
Skeletal wackestone-packstone	Burrow mottling	Fossil shells and skeletal debris	Mollusks (snails, clams), ostracods	Dolomite*, poikilotopic and anhydrite	Low to moderate; intercrystalline, vuggy	Inner ramp, middle ramp,	Cycle base
Oncoid wackestone-packstone	Burrow mottling	Algal oncoids, peloids, coated grains	Fusulinids	Dolomite	Low; intercrystalline	Middle ramp, outer ramp	Cycle base
Siltstone-sandstone	Burrow mottling	Detrital quartz	Rare mollusks	Quartz*, dolomite	Low; intercrystalline	Inner ramp, middle ramp, outer ramp	Cycle base, sequence base
Lithoclast wackestone		Sediment lithoclasts	Rare mollusks	Dolomite*, anhydrite	Low; intercrystalline	Tidal flat	Cycle top

**Signifies dominance.*

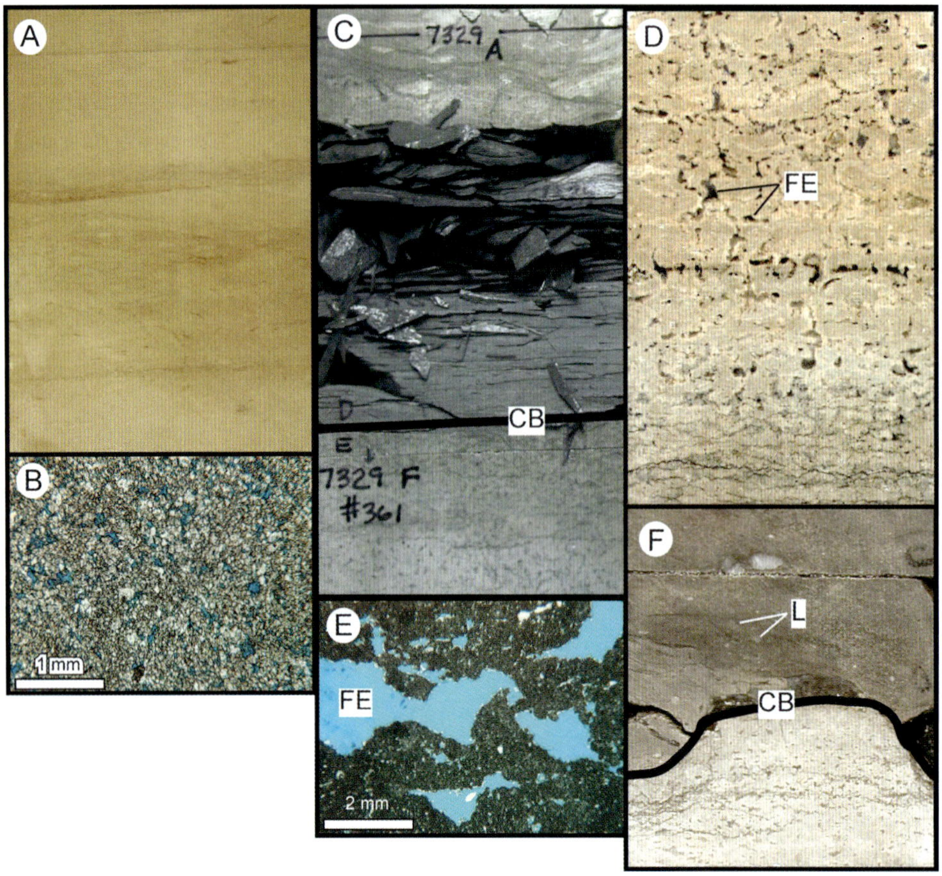

FIGURE 9. Core and thin-section photographs of typical tidal-flat facies in the Wichita and Lower Clear Fork. (A) Slab photograph of peritidal mudstone-wackestone showing weak laminations and local burrowing. The core is 10 cm (4 in.) wide. FCU 5927; depth, 2140 m (7021 ft). (B) Photomicrographs of peloid wackestone-packstone in A. Note the moldic pores. (C) Slab photograph of clay-rich carbonate mudstone facies in the Wichita. The core is 10 cm (4 in.) wide. Pan-American FM-1; well depth, 2234 m (7329 ft). Note the cycle boundary (CB) and underlying mudstone. (D) Slab photograph of exposed tidal-flat facies showing typical parallel laminations and fenestral pores (FE). The core is 10 cm (4 in.) wide. Lower Clear Fork FCU 7322; depth, 2075 m (6808 ft). (E) Photomicrograph of exposed tidal-flat facies showing fenestral vuggy pores (FE). FCU 6122; depth, 2104 m (6903 ft). (F) Slab photograph of CB showing fenestral-exposed tidal-flat facies overlain by subtidal, burrowed, peloid wackestone with skeletal debris. Note small lithoclasts (L) above the cycle top. Lower Clear Fork, top high-frequency sequence (HFS) L2.2. The core is 10 cm (4 in.) wide. FCU 7322; depth, 2074 m (6804 ft). FCU = Fullerton Clear Fork Unit.

Bebout, 2001), suggesting that these features are an expected and perhaps predictable feature of late highstand sedimentation and early diagenesis on Permian shallow-water carbonate platforms.

Peloid Packstone Facies

The peloid wackestone facies contain abundant peloids (Figure 10C) that are mostly fecal pellets created by burrowing infauna, although skeletal debris (chiefly mollusk fragments) are locally common. Nearly all low-energy mud-dominated facies in the Permian contain obvious pellets, indicating that burrowers were ubiquitous. A complete gradation from poorly defined peloids to identifiable pellets in textures ranging from mudstone to packstone exists. Differences in texture are probably caused by differences in early diagenesis and pellet preservation. Pellets that were lithified and stabilized early are more likely to be preserved. Thus, peloid (pellet) packstones are sediments that underwent significant amounts of early diagenesis, whereas peloid wackestones and pelleted mudstones underwent relatively little. The textural significance of these pelleted facies is thus more diagenetic than depositional. Accordingly, apparent variations in peloid abundance in these mud-dominated facies packstones, wackestones, and mudstones are not necessarily an indication of differing depositional environment or fluctuations in wave energy. Instead, they may be predominantly caused by local changes in rates and effects of early diagenesis. Porosity in these rocks

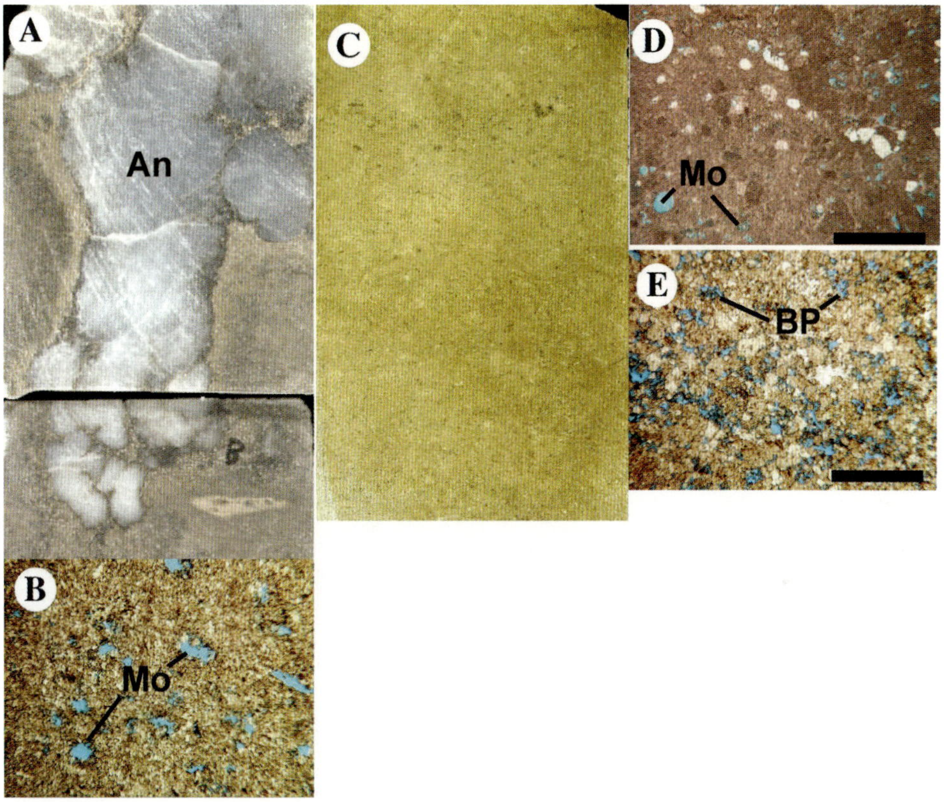

FIGURE 10. Core and thin-section photographs of typical peloid wackestone mud-rich peloid packstone-wackestone facies in Lower Clear Fork. (A) Slab photograph of peloid wackestone with anhydrite-bearing (An), solution-widened vertical burrow. Lower Clear Fork, HFS L 2.2. The core is 10 cm (4 in.) wide. FCU 7322; depth, 6874 ft (2095 m). (B) Thin-section photomicrograph of peloid wackestone with moldic pores (Mo). FCU 6229; depth, 6841 ft (2085 m). Scale bar = 1 mm. (C) Slab photograph of Wichita peloid packstone (dolostone). FCU 5927; depth, 7038 ft (2145 m). The core is 10 cm (4 in.) wide. (D) Photomicrograph of Lower Clear Fork peloid packstone (limestone) with moldic pores (Mo). FCU 5927; depth 6903 ft (2104 m). Scale bar = 1 mm. (E) Photomicrograph of Lower Clear Fork peloid packstone (limestone) with interparticle (BP) pores. FCU 6229; depth, 6817 ft (2078 m). Scale bar = 1 mm. FCU = Fullerton Clear Fork Unit.

is commonly associated with intercrystalline pores and less common skeletal moldic pores (Figure 10D, E).

Peloid Grain-Dominated Packstone Facies

Unlike the mud-rich peloid facies previously discussed, grain-dominated (or grain-rich) peloid packstones commonly display evidence of wave-related transport. These rocks are typically well sorted and contain interparticle pores that are either open or filled with cements (Figure 11A–D). The interparticle pores indicate that these peloids acted as true grains instead of pelleted mud, in contrast to the intercrystalline and moldic pores that typify mud-dominated facies. However, note that, in some instances, the grain-dominated texture is associated with vertical burrows and probably owes its origin to burrow-related diagenesis. This facies is restricted mostly to the subtidal legs of Lower Clear Fork depositional sequences. These rocks are among the highest quality reservoir facies in the field, with porosities as high as 20%.

Ooid-Peloid Grain-Dominated Packstone-Grainstone Facies

Ooid-peloid grain-dominated packstone-grainstone facies differ from the peloid grain-dominated packstone facies, with which they are commonly closely associated, in having recognizable ooids (as much as 250 μm in diameter) in addition to pervasive pellets (Figure 11E–G). Skeletal grains in the form of fusulinids, mollusks, and crinoids are also common. In most cases, these deposits, which are restricted to the subtidal sections of the Lower Clear Fork, are also well sorted. Grainstones, although relatively rare, display excellent size sorting and, in some cases, possess inclined or cross-laminations (Figure 11E). These rocks, which occur as both dolostone and limestone, represent the highest energy facies in the reservoir succession and, in many cases, display the best porosity and permeability. Pores are dominantly interparticle, but moldic pores are abundant, especially in limestone-dominated intervals (Figure 11B).

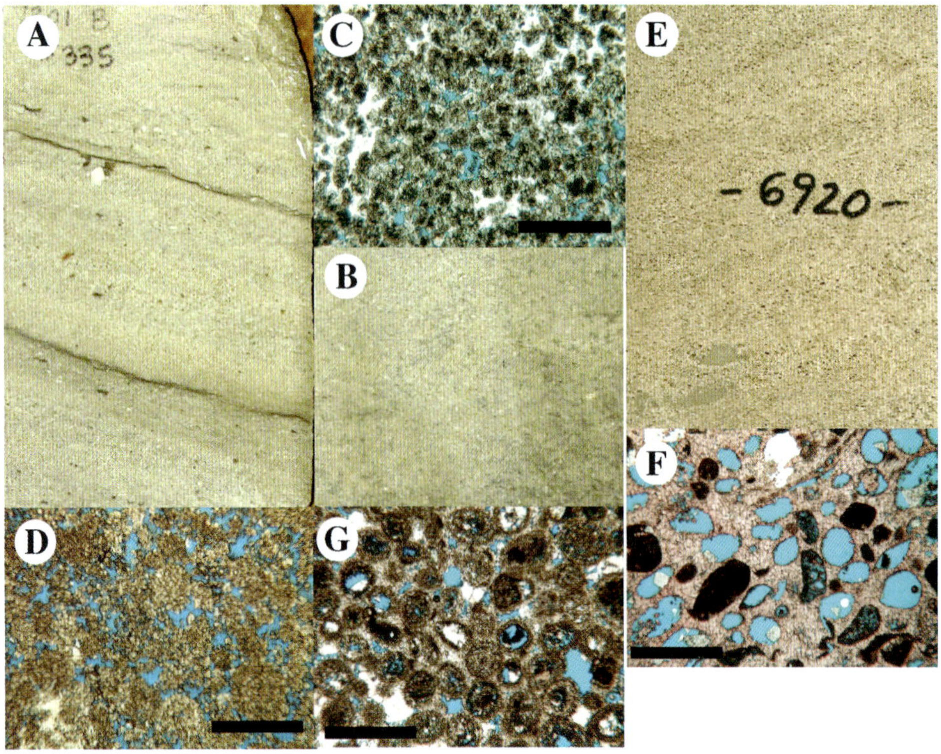

FIGURE 11. Core and thin-section photomicrographs of peloid grain-rich packstones and grainstones. (A) Slab photograph of Abo skeletal grain-dominated packstone showing inclined laminations. FCU 7322; depth, 7301 ft (2225 m). Core is 10 cm (4 in.) wide. (B) Slab photograph of Lower Clear Fork (L2.2) peloid grain-dominated packstone (dolostone). FCU 6229; depth, 6822 ft (2079 m). The core is 10 cm (4 in.) wide. (C) Photomicrograph of Lower Clear Fork peloid dolostone, grain-dominated packstone showing interparticle porosity. White areas are poikilotopic anhydrite. FCU 6122; depth, 6976 ft (2126 m). Scale bar = 1 mm. (D) Photomicrograph of Lower Clear Fork peloid grain-dominated packstone dolostone showing interparticle porosity. FCU 6229; depth, 6816 ft (2078 m). Scale bar = 1 mm. (E) Slab photograph of Lower Clear Fork ooid grainstone showing cross-laminations. FCU 6122; depth, 6920 ft (2109 m). The core is 10 cm (4 in.) wide. (F) Photomicrograph of Lower Clear Fork ooid limestone grainstone showing oomoldic pores. FCU 5927; depth: 6913 ft (2107 m). Scale bar = 1 mm. G. Photomicrograph of Lower Clear Fork ooid dolostone, grainstone, wackestone, showing interparticle and minor moldic pores. FCU 4828; depth, 7137 ft (2175 m). Scale bar = 1 mm. FCU = Fullerton Clear Fork Unit.

Fusulinid Wackestone-Packstone Facies

Fusulinid-bearing rocks are found in all three formations (Abo, Wichita, and Lower Clear Fork), although their presence is limited in the Wichita. In all occurrences, they are most commonly dolomitized. Fusulinids, which are preserved either as open or anhydrite-filled molds or as well-preserved fossil tests, can constitute as much as 40% of the rock (Figure 12A–D). In the Lower Clear Fork and Wichita, they are associated with abundant peloids (probably fecal pellets). In the Abo, they commonly co-occur with crinoids and less commonly with brachiopods. Fusulinids are thought to have occupied water depths of 98 ft (30 m) or more (Kerans and Fitchen, 1995) and represent the deepest water facies observed at Fullerton field. Accordingly, their presence is an indicator of platform flooding and relative sea level rise, making them key-indicator facies of cycle and sequence boundaries.

Skeletal Wackestone-Packstone Facies

Typically, skeletal wackestone-packstone rocks contain small volumes of skeletal debris (mostly mollusk fragments, but also crinoids and ostracods) and ubiquitous peloids. They grade into peloid wackestones and packstones. Evidence of burrowing is common. Typically, they contain low porosity; pore space consists of skeletal molds and intercrystalline pores. The dominance of mollusks and the essential absence of more normal marine organisms in these rocks suggest that they were deposited in an inner-platform setting.

Oncoid Wackestone-Packstone Facies

Oncoids (or oncolites) are large microbial-coated grains (Figure 12E–F) formed under conditions of continuous wave agitation in shallow water. They are abundant at the base of the Lower Clear Fork throughout the entire field area. Invariably, they are associated

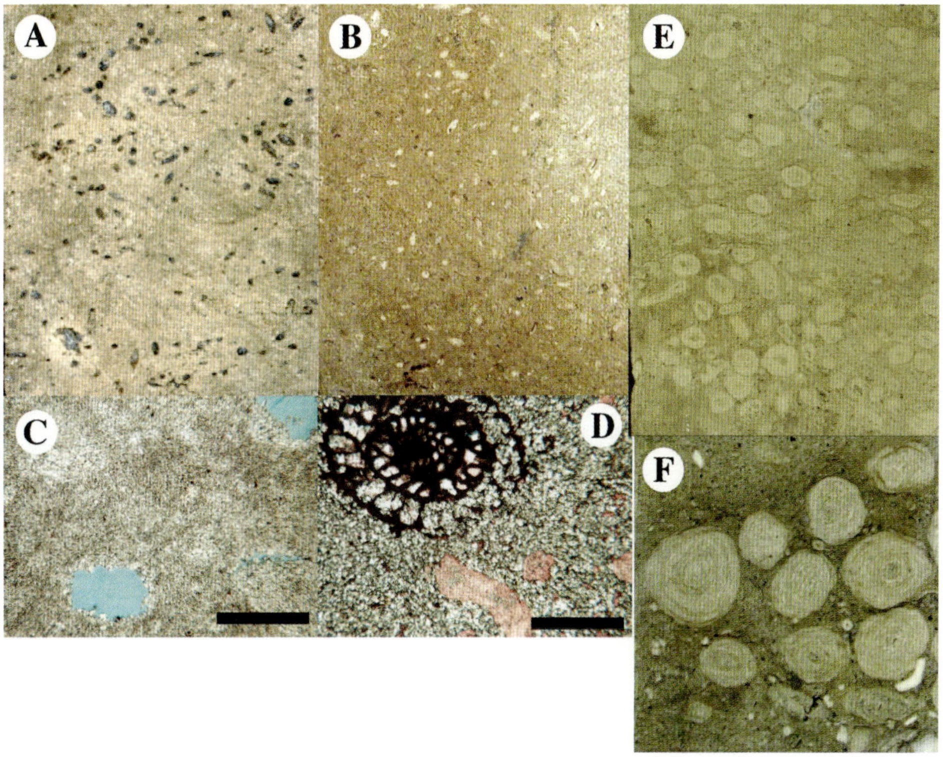

FIGURE 12. Core and thin-section photomicrographs of cycle base fusulinid wackestone-packstone and oncoid wackestone-packstone facies. (A) Slab photograph of fusulinid wackestone with open and anhydrite-filled fusumoldic pores. FCU 7322; depth, 6996 ft (2132 m). The core is 10 cm (4 in.) wide. (B) Slab photograph of fusulinid wackestone with anhydrite-filled fusumolds. FCU 7322; depth, 6994 ft (2132 m). The core is 10 cm (4 in.) wide. (C) Photomicrograph of fusulinid wackestone showing open fusumolds. FCU 7630; depth, 6778 ft (2066 m). Scale bar = 1 mm. (D) Photomicrograph of fusulinid wackestone showing well-preserved fusulinids but little or no porosity. FCU 4828; depth, 7018 ft (2139 m). Scale bar = 1 mm. (E) Small algally coated fusulinids. Lower Clear Fork FCU 5927; depth, 6980 ft (2128 m). (F) Large oncoids. Lower Clear Fork FM-1 well; depth, 7160 ft (2182 m). The core is 10 cm (4 in.) wide. FCU = Fullerton Clear Fork Unit.

with fusulinids and other faunas indicative of open-marine deposition. This association and their stratigraphic position immediately above the top of the tidal-flat-dominated Wichita indicate that they represent marine flooding of the platform during sea level rise (transgression). In some downdip wells, they are present at many intervals in the Lower Clear Fork, primarily at cycle bases. Their distribution suggests that they document relatively high-energy conditions developed during platform flooding. Like the fusulinid facies, the oncoid wackestone-packstone facies is a marker facies indicative of transgression or platform deepening (sea level rise).

Siltstone-Sandstone Facies

Quartz silt- and sand-bearing rocks are common only at the top of the Lower Clear Fork and in the overlying Tubb Formation. Because the quartz is associated with potassium- and thorium-rich clay minerals, its presence in the Lower Clear Fork and Tubb is well defined by high corrected gamma-ray wireline-log response (Figures 3, 7). The grain size of these rocks is fine sand to coarse silt, and they are generally subangular in shape, suggesting that these sediments were originally windblown. Quartz silt-sand occurs in two scenarios: in peritidal tidal-flat facies and in reworked subtidal facies. Peritidal tidal-flat facies were probably formed when small volumes of silt, sand, and clay were blown onto intermittently exposed, slowly accumulating tidal flats and admixed with peritidal carbonate sediment. The presence of potassium and thorium in these clastics produces the increased gamma-ray response that is associated with many tidal-flat deposits in the Clear Fork and facilitates the recognition of these cycle-capping deposits in these sediments (Ruppel, 2002). Most of the occurrences of silt, sand, and clay in the Lower Clear Fork at Fullerton are of this type.

Subtidal silt, sand, and clay deposits are most common in the Tubb. These rocks were formed first by large-volume eolian deposition during extended sea level lowstand and carbonate nondeposition and then

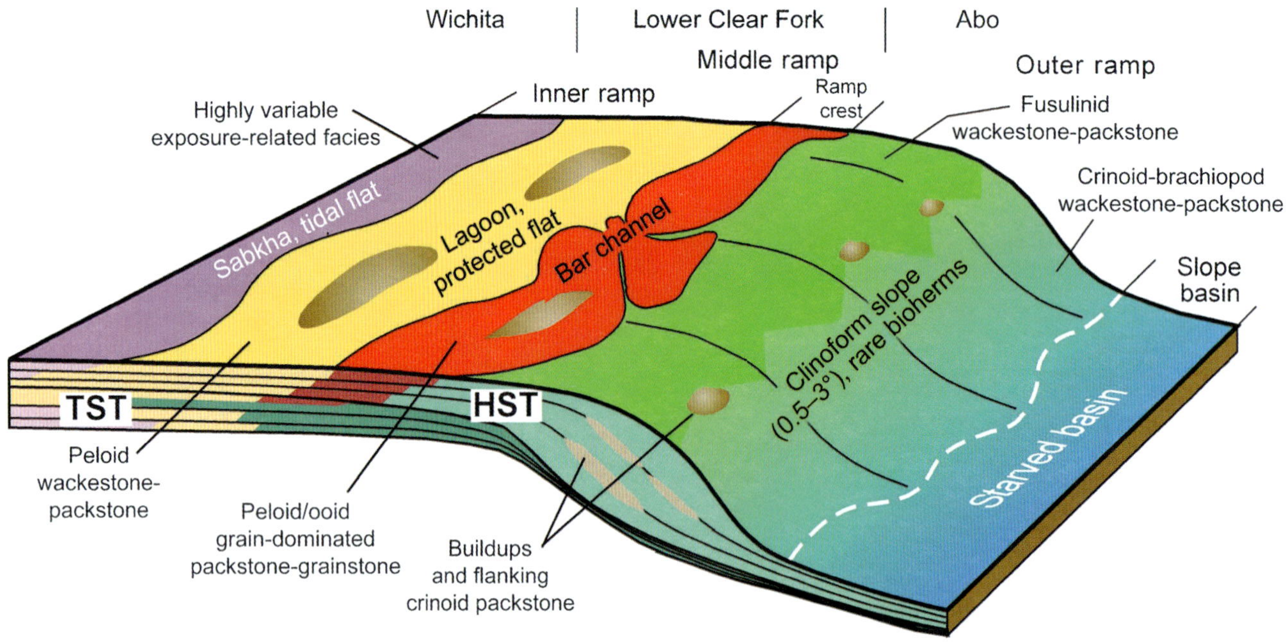

FIGURE 13. Depositional model for Permian shallow-water carbonate platforms in Permian Basin. Model applicable to most Leonardian and Guadalupian carbonate-platform successions, including reservoir succession at Fullerton field. Modified from Kerans and Ruppel (1994). TST = transgressive systems tract; HST = high standard systems tract.

reworked during the ensuing sea level rise and marine flooding of the platform. By contrast to the siliciclastics in the Lower Clear Fork, these deposits are generally richer in clastic content and intermixed with carbonate mud, and they show evidence of subtidal conditions (e.g., burrows, stratification). Although these rocks may locally exhibit some minor porosity, they do not appear to display any significant reservoir permeability.

Lithoclast Wackestone Facies

Thin intervals (commonly <1 ft [<0.3 m]) containing scattered lithoclasts are locally encountered at the contacts between transgressive marine facies and underlying tidal-flat deposits or other facies showing evidence of subaerial exposure (Figure 9F). Clasts are variable in composition but most commonly consist of fragments or intraclasts derived from the underlying bed. Typically, clasts are 0.4 in. (1 cm) in width or less. Although volumetrically minor, these rocks are an important indicator facies of lithification because of exposure or nondeposition and subsequent sea level rise and, thus, of cycle boundaries.

Depositional Model

The relative distribution of the facies previously described can best be understood when considered in light of a conceptual geologic model. Figure 13 portrays idealized 3-D relationships among major facies types and depositional environments typical of most middle Permian (Leonardian and Guadalupian) platform-carbonate successions in the Permian Basin. Note that this model reflects only relative interrelationships among facies tracts. The actual position of individual facies tracts at a given point in time is a function of many factors, including accommodation, rates and magnitude of sea level rise, rising versus falling sea level trends, climate, tectonics, and so on. During normal relative rises in sea level, facies tracts normally step landward; whereas during falls, they step basinward. The dominant depositional setting and facies of the Abo, Wichita, and Lower Clear Fork are depicted in Figure 13.

Sequence Stratigraphy

Figure 6 diagrammatically illustrates the sequence architecture, basic facies tracts interrelationships, and rock and time terminology of the lower Leonardian at Fullerton field. The parts of two composite or third-order depositional sequences and three facies tracts are represented in the field area. Four higher frequency sequences are also definable. The boundaries of these HFSs demarcate important correlative surfaces that are critical to defining architecture.

Sequence L1

The reservoir succession at Fullerton contains only the highstand part of sequence L1. In distal or basinward parts of the field area, sequence L1 is represented by Abo facies; whereas in updip landward areas, Wichita facies are dominant (Figure 6).

Abo Formation

Regionally, the Abo consists of outer-ramp-to-slope skeletal crinoid-fusulinid-dominated subtidal facies and less common, ramp-crest, ooid-peloidal grainstones (Kerans et al., 2000). In both outcrop and in the subsurface, the Abo is dominantly characterized by clinoformal top-lapping geometries (Kerans et al., 2000; Zeng and Kerans, 2003). This typical Abo architecture is apparent in 3-D and 2-D data from Fullerton field. The Abo is readily defined on seismic lines by its clinoformal reflectors and a prominent toplap surface that is apparent throughout the field (Figure 8).

Cores containing the Abo succession at Fullerton field display facies and bedding characteristics typical of sediments deposited in an outer-ramp-to-slope setting. Facies consist of alternating beds of fusulinid-crinoid packstones and wackestones and peloidal packstones. No true cyclicity is apparent in the Abo, although alternations exist between skeletal-rich and skeletal-poor intervals. Inclined beds are locally common. In addition, correlations within the Abo are not apparent either from core sections or from wireline logs partly because of limited core and well-log control through the Abo section. Such poor correlatability is typical of outer-platform depositional successions that display clinoformal architectures because of associated dipping bedding surfaces, point sources, discontinuous facies packages, and poor vertical facies contrasts.

Abo rocks are almost entirely dolomitized at Fullerton field. Porosity is locally high, reaching values of as much as 25%. Pore types vary among fusumolds, intercrystalline pores within coarse-crystalline dolomite, and interparticle pores in skeletal-peloidal packstones.

The thickness of the Abo is indeterminate at Fullerton for two reasons. First, few wells and no cores penetrate the complete Abo to the underlying Wolfcamp section. Second, outcrop studies reveal that the Abo-Wolfcamp contact may be lithologically indistinct in many areas. In outcrop, both the uppermost Wolfcamp and the Abo are composed of clinoformal, outer-platform, fusulinid-crinoid wackestones. A minimum Abo thickness of 300 ft (91 m) is established by a cored well (Amoco FM-1) in the southeast corner of the field (Figure 4).

The position of the L1-L2 sequence boundary is readily apparent in downdip areas, where top L1 Abo facies are karsted and sharply overlain by basal tidal-flat facies of the basal transgressive L2 Wichita. In these areas, the sequence boundary is marked by an interval (up to several feet thick) of intermixed clasts and fragments of Wichita tidal-flat deposits and Abo fusulinid-bearing subtidal deposits.

Wichita Group

The top of sequence L1 consists of Wichita Group facies in more updip or landward (westward) areas of Fullerton field (Figure 6). The Wichita consists of a diverse assemblage of peritidal to supratidal tidal-flat deposits. Outcrop and subsurface studies demonstrate that the Wichita comprises parts of two depositional sequences: the highstand leg of sequence L1 and the transgressive leg of sequence L2 (Figure 6). The lower part of the Wichita Formation is an updip proximal equivalent of Abo outer-platform succession of sequence L1, whereas the upper Wichita is a proximal equivalent of Lower Clear Fork subtidal outer-platform rocks of sequence L2.

The L1 highstand Wichita deposits are pervasively dolomitized. Facies are dominantly structureless peritidal mudstone-wackestones and clay-rich carbonate mudstones (Figure 9A, C). The L1 Wichita rocks, like the total Wichita succession, thin eastward as they grade into subtidal Abo facies (Figures 14A, 15). Although not readily definable on wireline logs, core studies suggest that basal Wichita rocks differ from overlying L1 and L2 Wichita deposits in containing fewer clay-rich carbonate mudstones (Figure 15). This is consistent with the basal Wichita representing deposition in a more distal tidal-flat setting than the upper part. Neither cycles nor HFSs are readily definable in the L1 Wichita because of the lack of facies contrast. Similarly, the L1-L2 sequence boundary is difficult to define in proximal areas of the field where Wichita facies of the L1 highstand systems tract (HST) are overlain by very similar L2 Wichita transgressive systems tract (TST) facies. However, the approximate position of this sequence boundary is suggested by the distribution of karst breccias. Although not present in all cores, they are found at approximately the same stratigraphic horizon in several wells (Figure 7).

Sequence L2

Most of the reservoir succession at Fullerton is assignable to sequence L2, which is subdivisible into four HFSs (Figures 6, 15).

High-Frequency Sequence L2.0

High-frequency sequence L2.0 documents initial flooding of the platform following sea level fall and

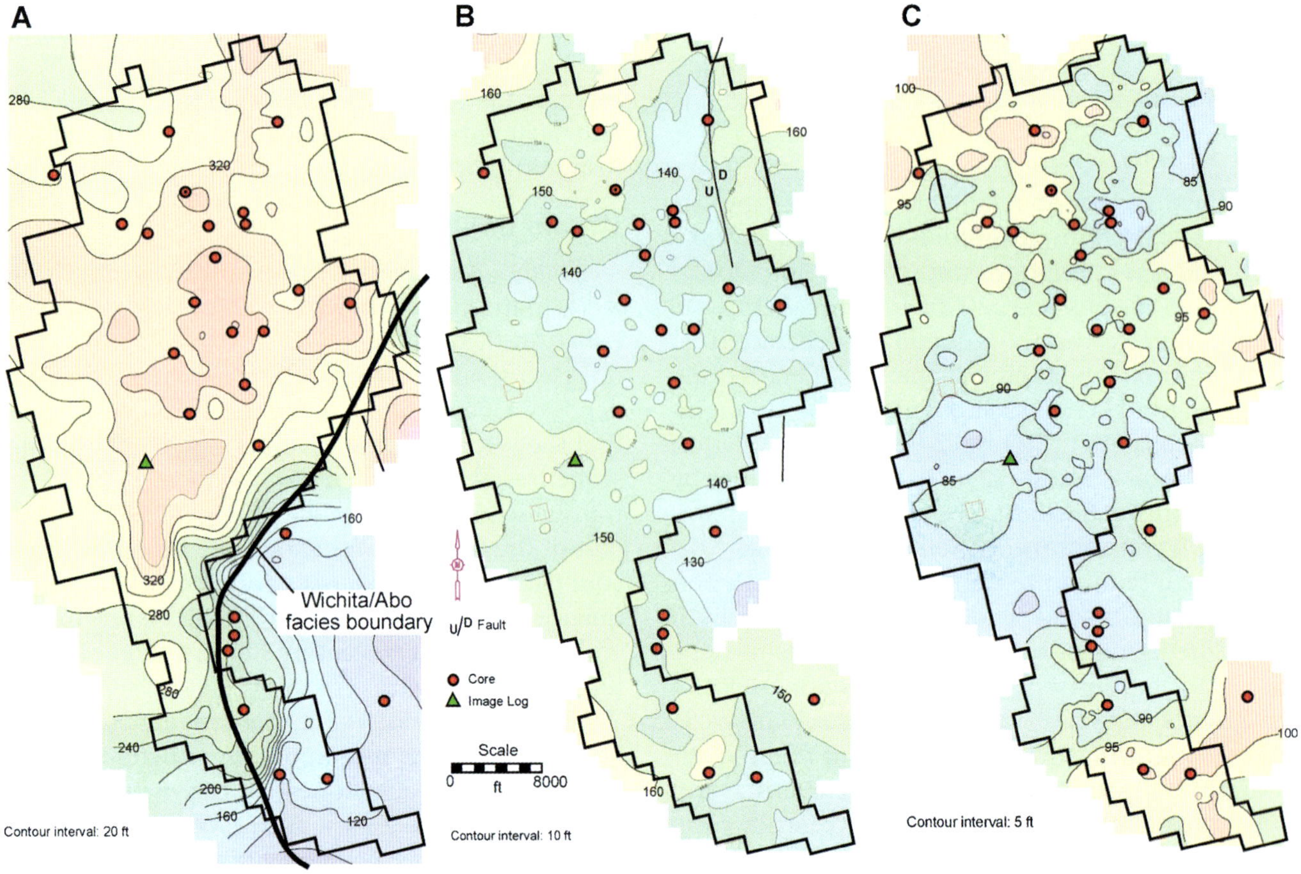

FIGURE 14. Thickness of Wichita and Lower Clear Fork successions in Fullerton field area. (A) Total Wichita thickness. The relatively sharp decrease in thickness in the southeastern part of the area is caused by the change in facies from Wichita tidal-flat facies to time-equivalent subtidal Abo facies. (B) Thickness of Lower Clear Fork L2.1 sequence. Note abrupt change in thickness along north-trending fault in the northeastern part of the field. (C) Thickness of Lower Clear Fork L2.2 sequence.

exposure at the end of L1 deposition. In most of the Fullerton field area, HFS L2.0 consists of amalgamated tidal-flat deposits of the upper Wichita. Subtidal Lower Clear Fork L2.0 deposits are present only at the downdip margins the field (Figures 6, 15). The thickness of the Wichita is relatively uniform over the northern two thirds of the field, ranging from about 350 ft (~107 m) in the center of the field to about 280 ft (~85 m) along the western, northern, and eastern margins of the field area (Figure 14A). The thickness decreases, however, to as little as 110 ft (34 m) in the southeastern part of the field (Figure 14A). This decrease in thickness, which occurs relatively abruptly along a generally northeast-trending belt, marks the facies change from the updip inner-platform Wichita tidal-flat facies to the downdip Abo outer-ramp fusulinid facies (Figure 15).

Correlations within the thick Wichita succession of relatively similar tidal-flat deposits are difficult because of (1) discontinuity of tidal-flat deposits, (2) lack of any systematic vertical stacking relationships among facies to define cycles, (3) nonuniform response of wireline logs to these facies, and (4) overprinting effects of diagenesis. The presence of several beds of limestone in the upper Wichita in the northern part of the field makes correlations much more robust in that area of the field (Figure 7). These limestone beds differ primarily from interbedded dolostone by being extremely low in porosity.

High-Frequency Sequence L2.1

In most of the Fullerton field area, HFS L2.1 defines the base of the Lower Clear Fork. High-frequency sequence L2.1 consists of a basal section of transgressive to early highstand subtidal platform facies and an upper section of late highstand tidal-flat facies (Figures 3, 7, 15). The basal transgressive subtidal facies of the Lower Clear Fork represent the first marine flooding of the platform and a sharp change in depositional style

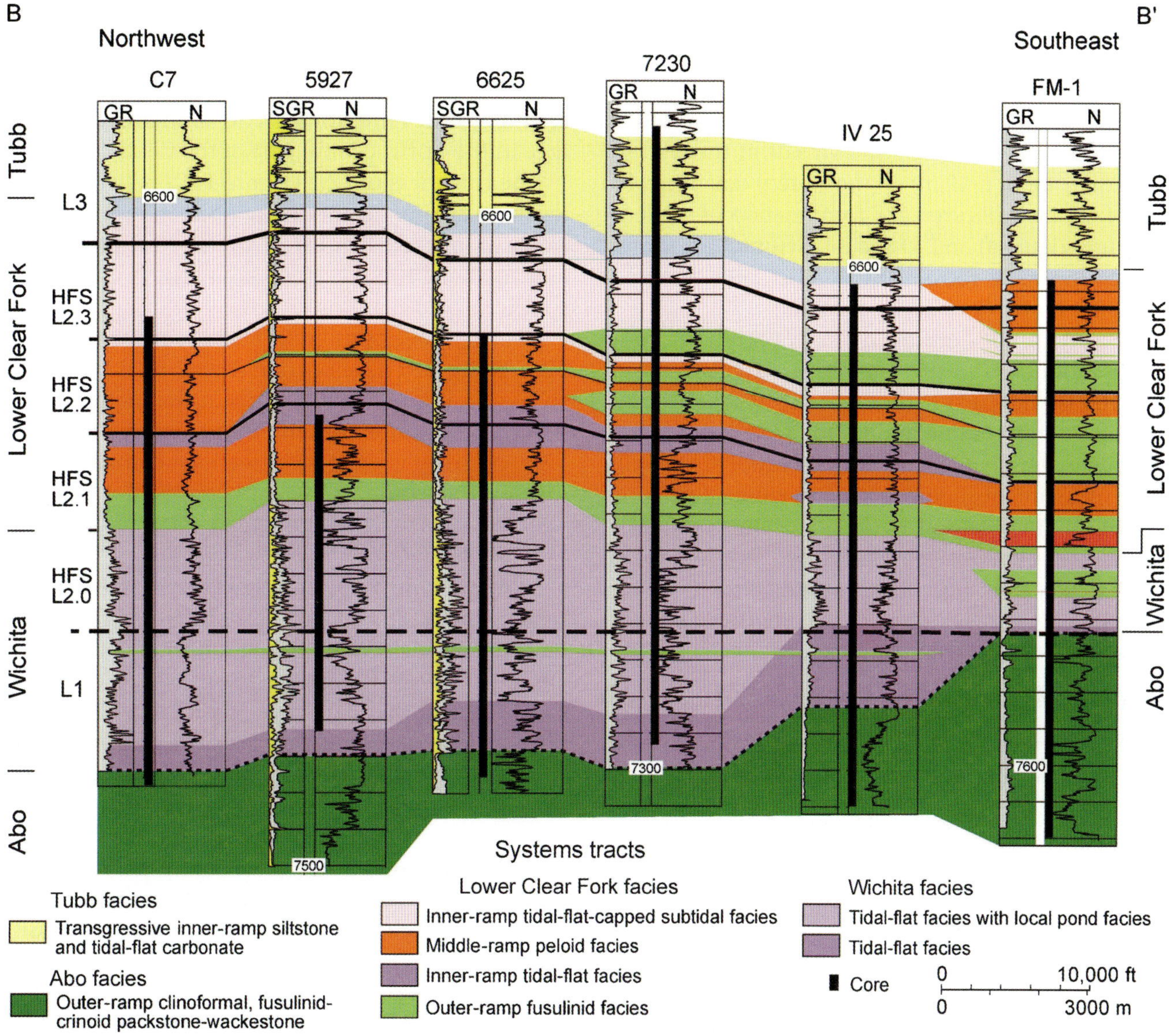

FIGURE 15. Northwest–southeast cross section (BB′) of the Fullerton field area showing the sequence architecture and general facies development based on cored well control. Line of section shown in Figure 4. HFS = high-frequency sequence; N = neutron log; GR = gamma ray; SGR = spectral gamma ray. From Ruppel and Jones (2006).

from tidal-flat deposition of the Wichita to subtidal deposition of the Lower Clear Fork.

The basal transgressive leg of L2.1 dominantly consists of fusulinid wackestone-packstone and oncoid wackestone-packstone facies that document landward backstepping of outer-platform facies across the platform. These rocks are generally overlain by an upward-shallowing succession of peloid packstones and grain-dominated packstones that represent late transgression and early highstand. Locally, within this succession, tidal-flat facies are exposed, indicating periodic exposure. High-frequency sequence L2.1 is capped throughout most of the field by a succession of one to three tidal-flat-capped cycles that represent exposure during late highstand (Figures 7, 15). These tidal-flat facies typically exhibit an elevated gamma-ray log response that aids in their recognition (Figures 3, 7, 15).

Like most of the Lower Clear Fork in the Permian Basin, HFS L2.1 is dominantly dolostone. However, limestone is locally common at Fullerton field, especially in the transgressive leg of L2.1 (Figure 3). Areally, limestone is most abundant along the periphery of the northern half of the field and at the southern end of the field (Harrington and Lucia, 2012).

The thickness of L2.1 is relatively constant across the Fullerton field area, ranging from about 140 to 150 ft (~43–46 m) across most of the area (Figure 14B). In some cases, thickness changes appear to be a function of differential subsidence along deep-seated faults. Thickness changes are especially apparent along the north-trending fault in the northeastern part of the field, where thickness changes from less than 140 ft (<43 m) on the western upthrown side of the fault to more than 160 ft (>49 m) on the eastern downthrown side of the fault (Figure 14B). Examination of modern structure (Figure 4) on 2-D and 3-D seismic data shows that considerable movement on this fault occurred even after Lower Clear Fork deposition.

Porosity can be developed in any part of L2.1 but is most abundant in the subtidal (transgressive and early highstand) parts of the sequence. Pores are dominantly intercrystalline and moldic; the latter being especially abundant in limestone sections. Overall, these rocks are most porous in areas where limestone is present. Tidal-flat rocks are also locally porous but generally contain low permeabilities. Subtidal rocks of L2.1 constitute the most productive reservoir interval in Fullerton field, as well as in many other fields across the Central Basin platform.

High-Frequency Sequence L2.2

High-frequency sequence L2.2 is similar to HFS L2.1 in consisting of a basal transgressive leg composed of backstepping tidal-flat facies, a middle (late transgressive to early highstand) leg composed dominantly of subtidal facies, and an uppermost (late highstand) leg composed of tidal-flat facies (Figures 7, 15). The basal L2.2 tidal-flat deposits are thickest in the center of the field but generally absent along the margins, reflecting greater accommodation and early flooding of downdip areas. The lower abundance of outer-ramp fusulinid-rich facies in the TST of L2.2 relative to L2.1 suggests that overall accommodation during L2.2 sea level rise was somewhat less than during L2.1, implying that maximum flooding for sequence L2 is represented by HFS 2.1. Lower accommodation during L2.2 and attendant lower wave energies are also suggested by the near absence of the oncoid wackestone-packstone facies in L2.2—these rocks are found only in the most downdip core in the field. The fusulinid wackestone-packstone facies is also mostly restricted to the eastern and southeastern parts of the area (Figure 15). Throughout most of the field area, L2.2 is dominated by peloid wackestones and packstones (Figures 7, 15) typical of middle-platform deposition. The sequence is capped by a thin succession of tidal-flat cycles (Figures 7, 15).

High-frequency sequence L2.2 is dominantly composed of dolostone with two important exceptions. As in L2.1, limestone is dominant in the southern part of the field (Harrington and Lucia, 2012). Dolomite in this area is mostly restricted to mud-rich fusulinid wackestones; virtually all grain-rich facies are limestone. Limestone is also abundant in a small area in the north-central part of the field.

The thickness of L2.2 (Figure 14C) is significantly less than that of L2.1, ranging from about 85 to 90 ft (~26–27 m) across most of the central part of the area to about 100 ft (~30 m) along the northern and southern field margins.

Porosity development in L2.2 is similar to that in L2.1, being best developed in the subtidal (transgressive and early highstand) parts of the sequence and dominated by intercrystalline and moldic pores. Like L2.1, porosity is greatest in sections containing significant limestone.

High-Frequency Sequence L2.3

The uppermost Lower Clear Fork high-frequency sequence (L2.3) is composed of tidal-flat-capped restricted subtidal cycles throughout most of the field area (Figures 7, 15). Log correlations suggest that these cycle-capping tidal-flat facies are relatively continuous across significant areas of the field, in contrast to those of the Wichita tidal-flat facies. The tops of these cycles are relatively easily correlated with porosity logs because of the typical development of fenestral and fine intercrystalline pores. The underlying cycle-base subtidal rocks are dominantly mud-rich packstones and wackestones. For the most part, porosity is restricted to the cycle-top tidal-flat caps. However, because the porosity in these rocks is dominated by fine intercrystalline and moldic pores, they exhibit very little permeability. A possible exception occurs along the outer margins of the field, where grain-rich peloidal packstones of the middle ramp become more common (Figure 15). If reservoir permeability is present anywhere in L2.3, it is in these field-margin areas.

Sequence L3

Although containing no reservoir facies, sequence L3 is important for correlation of underlying reservoir intervals. Regionally, sequence L3 is characterized by basal TST deposits of Tubb Formation siltstones and silty dolostones. At Fullerton, the L2-L3 sequence boundary is operationally picked at correlative beds of silty peritidal facies that underlie a widespread interval of fusulinid wackestones (Figure 15). This flooding succession is overlain by thicker beds of Tubb siltstones and silty dolostones, which are relatively easily defined by a high gamma-ray log response (Figures 3, 15). Tubb siliciclastics are interpreted as products of eolian deposition that accumulated during the post-L3 lowstand and that were reworked during the ensuing L3 sea level transgression. Because of this reworking, most Tubb intervals contain varying mixtures of siltstone-sandstone and carbonate (typically mud-rich shallow-water facies). Although some Tubb beds appear correlative locally, they are highly variable in their distribution across the field area.

Cycle-Scale Stratigraphy

The fundamental goal of cycle stratigraphy is to develop a correlation framework based on time-equivalent surfaces (e.g., cycle tops). The basic underlying premise for this approach is the assumption that widely correlative depositional cycles are formed by punctuated allocyclic processes (e.g., sea level rise and fall) that affected sedimentation over broad areas. The methodology used to develop a cycle-scale stratigraphic framework in the Leonardian section at Fullerton field consists of (1) characterization of facies-stacking patterns and cycle development in analogous outcrops; (2) description, interpretation, and logging of facies-stacking patterns and possible cycle tops in cores; (3) integrated log- and core-based correlation of tentative cycle tops; and (4) definition of cycle architecture. At Fullerton field, nearly 15,000 ft (4572 m) of core (from 29 cored wells) was described in detail to provide the basic data for cycle definition.

Wichita Cyclicity: L1 and High-Frequency Sequence L2.0

Tidal-flat facies, such as those that characterize the Wichita, are typically highly discontinuous laterally; depositional cycles are thus generally not easily definable on the basis of facies-stacking relationships in these rocks. This assertion is supported both by outcrop studies (e.g., Ruppel et al., 2000) and by examination of such sediments in modern settings. Only one widely correlative cycle top was defined in the Wichita on the basis of facies-stacking relationships. This surface, which occurs in about the middle of the Wichita succession in most of the field, is marked by the sharp superposition of clearly subtidal rocks over the more typical tidal-flat facies of the Wichita. The base of this cycle is marked by a low gamma-ray log response (Figures 7, 15) associated with subtidal fusulinid-bearing facies and the top by a high gamma-ray cycle-top tidal-flat facies (peritidal mudstone-wackestone facies or exposed tidal-flat facies).

As described above, cyclicity, also definable in the upper part of the Wichita, can be defined by log response in some parts of the field based on mineralogy trends. These cycles are composed of high-porosity dolostone cycle tops and low-porosity limestone cycle bases. Because of the contrast in both mineralogy and porosity, both cycle-base and cycle-top facies can be defined readily from porosity logs (neutron and density logs). The average of these cycles ranges from about 10 to 25 ft (~3–7 m) in thickness (see Lucia and Kane, 2012).

Lower Clear Fork Cyclicity: High-Frequency Sequence L2.1 and L2.2

Lower Clear Fork cycles are typically characterized by having mud-rich facies at their bases and grain-rich facies at their tops. Basal transgressive cycles in HFS L2.1 contain abundant oncoids at cycle bases, along with accompanying fusulinids (Figure 16A). These cycles are capped by better sorted peloid-rich facies. High-frequency cycles (typically 5–10 ft [2–3 m] in thickness) stack into cycle sets that average 25 to 40 ft (7.62 m to 12.19 m) in thickness (Figure 16A). Cycle sets display facies-stacking patterns similar to cycles, consisting of fusulinid-rich bases and peloid-rich caps. Tidal-flat facies cap these cycle sets locally. Porosity is generally highest at both cycle tops and cycle-set tops relative to bases. Because cycles in the lower parts of cycle sets are generally more mud and fusulinid rich, however, the overall porosity in these basal TST cycles is generally relatively low.

Cores and outcrop studies suggest that cycles and cycle sets are correlative over significant distances (e.g., Ruppel et al., 2000). However, where cores are not available, these correlations can be difficult to establish. Because gamma-ray logs display virtually no systematic response to subtidal facies or cycles (Figure 16A, B), they cannot be used reliably for cycle-scale correlation in most of the Lower Clear Fork. In the absence of gamma-ray logs, the most effective way of establishing cycle-scale correlations is through the use of porosity logs. This approach is based on the observation from outcrops and cores that cycle tops consistently contain the most grain-rich facies and that these rocks are most likely to contain high porosity. Because of this relationship, porosity logs can be used to correlate both high-porosity facies in the upper part of the cycle and the overlying cycle top.

The L2.1 highstand cycles are dominated by mud-rich peloidal facies (Figure 10C–E) at their bases and grain-rich peloid- or ooid-bearing facies at their tops (Figures 11B–G, 16B). Fusulinids are mostly uncommon in these highstand cycles, reflecting the basinward shift in facies tracts. In general, these cycles are dominated by peloid packstones and grain-rich packstones.

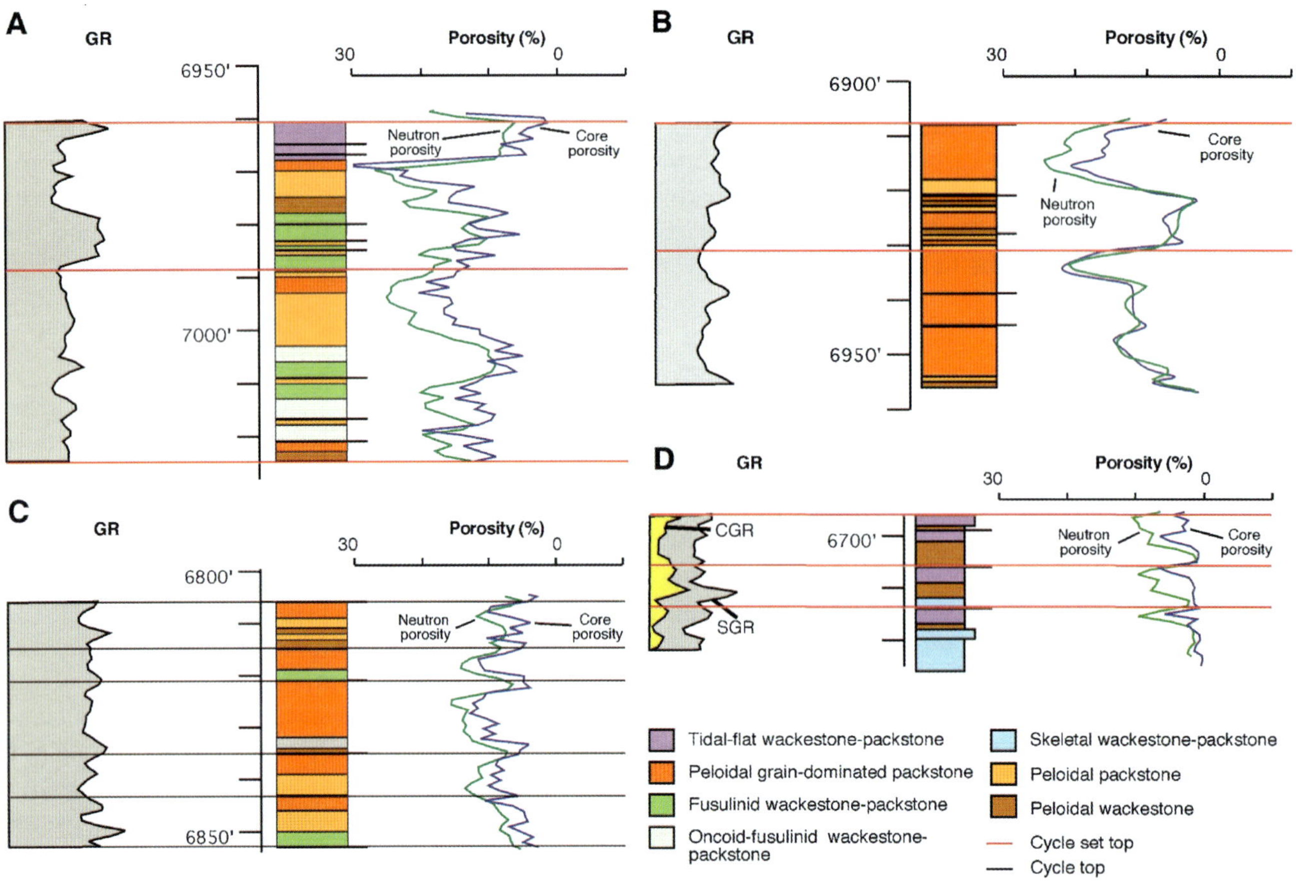

FIGURE 16. Styles of facies stacking and cycle development in the Lower Clear Fork. (A) Fusulinid- and oncoid-rich transgressive systems tract (TST) of the Lower Clear Fork high-frequency sequence (HFS) L2.1. Note that porosity is typically developed at or near cycle tops. (B) Grain-rich highstand systems tract (HST) of Lower Clear Fork HFS L2.1. Note that porosity is typically developed at or near cycle tops. (C) Grain-rich late TST–early HST of Lower Clear Fork HFS L2.2. Note that porosity is typically developed at or near cycle tops. (D) Cycle-capping tidal-flat facies at cycle tops, but little or no permeability is associated with fenestral pores that dominate these caps. Note that with all cycle types, no systematic relationship exists between gamma-ray log response and facies or cyclicity. Lower Clear Fork HFS L2.3. SGR = spectral gamma ray; CGR = corrected gamma ray; GR = gamma ray. Modified from Ruppel and Jones (2006).

Porosity is highest at both cycle tops and cycle-set tops because of the abundance of these grain-rich facies. Cycles typically average 5 to 10 ft (2–3 m) in thickness; cycle sets are commonly 20 to 30 ft (6–9 m) thick (Figure 16B).

Facies stacking and cycle development in HFS L2.2 are similar to those in L2.1. Cycles average 5 to 10 ft (2–3 m) in thickness, and porosity is best developed in cycle tops (Figure 16C). The L2.1 cycles differ, however, in the lack of oncoid facies and the relative scarcity of fusulinid facies, presumably reflecting decreasing overall accommodation in L2.2 because of platform aggradation and slowing rates of long-term sea level rise. In addition, cycle sets are not generally definable in L2.2 mostly because of the general absence of fusulinid facies that define these intermediate-scale sea level rise events in L2.1.

Lower Clear Fork Cyclicity: High-Frequency Sequence L2.3

The high-frequency cyclicity in the Leonardian at Fullerton field is most readily definable in L2.3. These rocks, which are characterized by tidal-flat-capped shallow subtidal cycles (Figure 16D), appear to be much more widely correlative than cycles in L2.1 or L2.2. Although neither facies nor cyclicity is defined by gamma-ray logs, both are distinguishable on porosity logs because of the typically well-developed porosity associated with cycle-capping tidal-flat deposits. Porosity in these rocks is generally caused by the presence of fenestral pores and can be relatively high, especially relative to the generally low porosity exhibited by cycle-base mud-rich wackestones and packstones. Because of the separate-vug fenestral pores, however, permeability is generally low, and L2.3 rarely contributes to hydrocarbon production in the field.

Porosity-based wireline correlations suggest that cycle and facies continuity is high. This result is somewhat unexpected, given outcrop observations suggesting that analogous tidal-flat facies are highly discontinuous. Laterally continuous porosity development at these cycle tops probably reflects the effects of early diagenesis associated with sea level fall and subsequent rise at these surfaces.

MINERALOGY AND DIAGENESIS

The lower Leonardian section at Fullerton field is similar to most other platform Leonardian successions in the Permian Basin in showing evidence of significant postdepositional diagenesis. The principal products of this diagenesis are matrix-replacive and pore-filling dolomite and anhydrite. Dolomite is dominant in Fullerton as it is in most Leonardian reservoirs in the Permian Basin. However, limestone is locally present at Fullerton in the Abo, Wichita, and Lower Clear Fork.

Dolomite and Limestone Distribution

Dolomite is by far the dominant mineral in the reservoir section. The Abo entirely consists of dolomite, except in the most downdip wells. In the core from the FM-1 well (Figure 4), the Abo contains alternating zones of limestone and dolostone. Dolostone is typically associated with mud-rich facies (peloid wackestone and fusulinid wackestone) in the Abo clinoformal succession, (Figure 15) whereas limestone intervals are more commonly grain-rich skeletal facies.

Calcite is locally abundant in the Wichita, especially in the upper half of the formation (Figure 7). In all cases, undolomitized rocks in the Wichita are peritidal mudstones or wackestones (peritidal mudstone-wackestone facies). These facies characteristically exhibit low porosity (<2%) and essentially no permeability. Log correlations suggest that limestone intervals are locally correlative (Figure 7).

Lower Clear Fork rocks at Fullerton field also contain locally abundant limestone, which is common in HFS L2.1 across most of the field (Figures 3, 7) (Harrington and Lucia, 2012). The highest abundance is in the southern part of the field, where only mud-rich (typically fusulinid-bearing) facies contain significant dolomite. Essentially all grain-rich facies in this area are dominantly undolomitized. In the northern part of the field, limestone is also common in L2.1 but is complexly distributed among all subtidal facies. Limestone is also abundant in L2.2 in the southern part of the field but rare in other parts of the field (Harrington and Lucia, 2012). Porosity is developed in both dolostone and limestone-rich intervals in the Lower Clear Fork.

Stable Isotope Chemistry

To constrain styles and timing of diagenesis and porosity evolution in the reservoir, we collected stable isotope data from several cores in the field (Figure 17). Data collected and reported by Ruppel and Jones (2006) were supplemented with 90 additional data points for this study. Three main conclusions can be drawn from these data: (1) limestones exhibit seawater calcite $\delta^{18}O$ values, indicating that they formed in seawater and have not been diagenetically altered; (2) dolostones exhibit spatial trends in $\delta^{18}O$ that suggest that they were formed during different episodes of dolomitization and by significantly different diagenetic fluids; and (3) $\delta^{13}C$ data define secular changes in seawater carbon budget that is persistent over large areas of the Permian Basin and that are useful for correlations.

Limestones

Limestones from the Wichita and Lower Clear Fork exhibit an average $\delta^{18}O$ value of −2.67‰ Peedee belemnite (PDB) (n = 13; range, −1.55 to −3.55‰; Figure 17). This range is consistent with the current best estimate for marine calcite precipitated from seawater during the middle Permian of −2.8‰ PDB (Lohman and Walker, 1989) and is also similar to values reported for Leonardian dolostones by Mazzullo (1994). The good agreement between Fullerton calcites and limestones to seawater $\delta^{18}O$ value suggests that these rocks contain a preserved record of original seawater chemistry and, by extension, that these rocks have undergone relatively little chemical alteration.

Dolostones

As a group, Fullerton dolostones display a fairly wide range of both $\delta^{18}O$ and $\delta^{13}C$ values (Figure 17). Carbon isotopes values mostly range from +2 to +5‰ PDB, which is similar to that observed by previous authors for Leonardian dolostones (Ye and Mazzullo, 1993; Mazzullo, 1994; Saller and Henderson, 1998; Ruppel, 2002). Oxygen isotopes fall in a somewhat broader range from about −1 to +3‰ PDB (Figure 17). Although similar in part to previously published data (Ye and Mazzullo, 1993; Saller and Henderson, 1998; Ruppel, 2002), Fullerton $\delta^{18}O$ values exhibit more of a spread toward depleted values. When examined

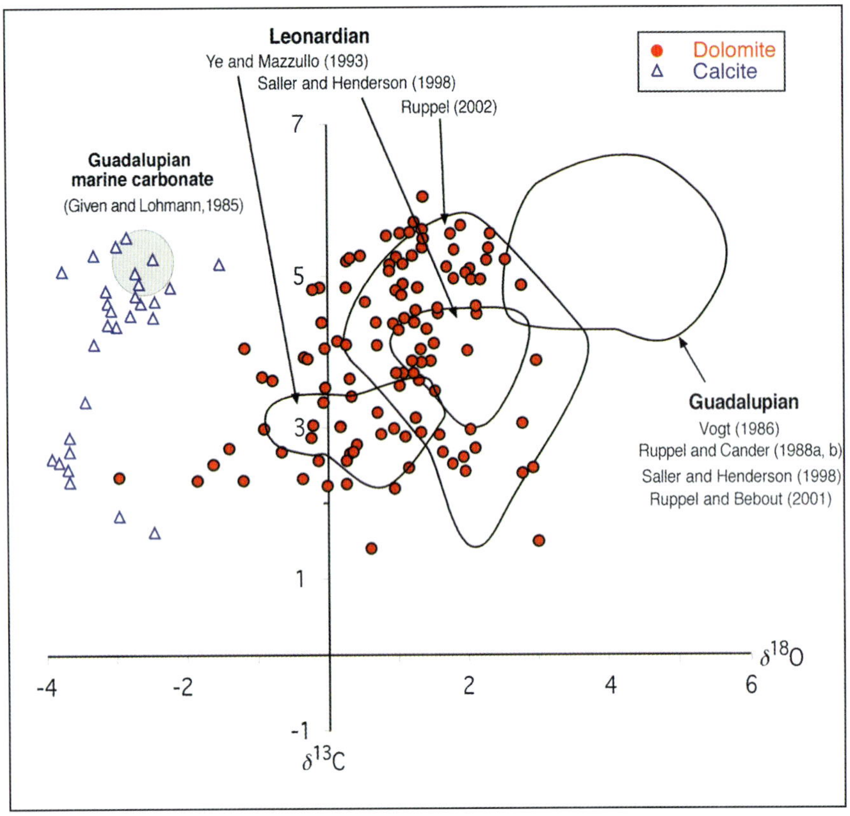

FIGURE 17. Stable isotope data for Tubb, Wichita, Lower Clear Fork, and Abo at Fullerton field. Solid red circles and blue triangles are dolomite and calcite samples, respectively, from Ruppel and Jones (2006). Open green circles and green triangles are dolostone and limestone samples, respectively, collected during this study. Modified from Ruppel and Jones (2006).

stratigraphically, however, both carbon and oxygen isotope data reveal systematic spatial trends that have importance for understanding porosity evolution and reservoir architecture.

The $\delta^{13}C$ data reveal a progressive change from lighter values (+2 to +4‰ PDB) in the Abo and Wichita (sequences L1 and L2.0) to heavier values (+4 to +6‰ PDB) in the upper Wichita and Lower Clear Fork (sequences L2.1 to L2.3). This change is observed in all cores sampled and suggests a systematic increase in seawater $\delta^{13}C$ during L2.0 time (Figure 18). This trend has also been observed in North Dollarhide field Leonardian cores 30 mi (48 km) to the southwest of Fullerton field (Ruppel, unpublished data). Relatively depleted values of $\delta^{13}C$ in the Wichita and Abo were also recorded by Ye and Mazzullo, (1993) (Figure 18), suggesting that this trend in $\delta^{13}C$ is widespread in the Permian Basin. The close agreement between positions of defined sequence boundaries in the field area and $\delta^{13}C$ trends also provides support for the accuracy of correlation-based sequence stratigraphy and the reservoir architecture that is based on it.

The $\delta^{18}O$ values at Fullerton are similar to previous Leonardian data reported by Ye and Mazzullo (1993), Saller and Henderson (1998), and Ruppel (2002) but are very different from most data reported for younger Guadalupian platform dolomites. Typical $\delta^{18}O$ values for these Guadalupian (San Andres and Grayburg formations) rocks are +3 to +6‰ PDB (Vogt, 1986; Ruppel and Cander, 1988a, b; Saller and Henderson, 1998; Ruppel and Bebout, 2001; Ruppel, 2002). The relatively enriched $\delta^{18}O$ signatures of Guadalupian dolomites have invariably been interpreted to have been produced during dolomitization by evaporatively concentrated seawater brines. The relatively depleted values for Leonardian dolomites at Fullerton and in other Leonardian fields suggest that (1) dolomitization was caused by brines that were less evaporatively concentrated than those that caused Guadalupian dolomitization, (2) dolomitization was caused by fluids of a mixed water origin, or (3) dolomitization was caused by a combination of both 1 and 2. Geographic and stratigraphic patterns of $\delta^{18}O$ data across the field also suggest relationships between depositional settings, structural position on the platform, early and late diagenesis, and porosity formation. These relationships are presented and discussed in the section Models for Porosity Development below.

Karst Development

Karst Fabrics

Sedimentary breccias are present in several cores. Although thin intervals of conglomerates and breccias

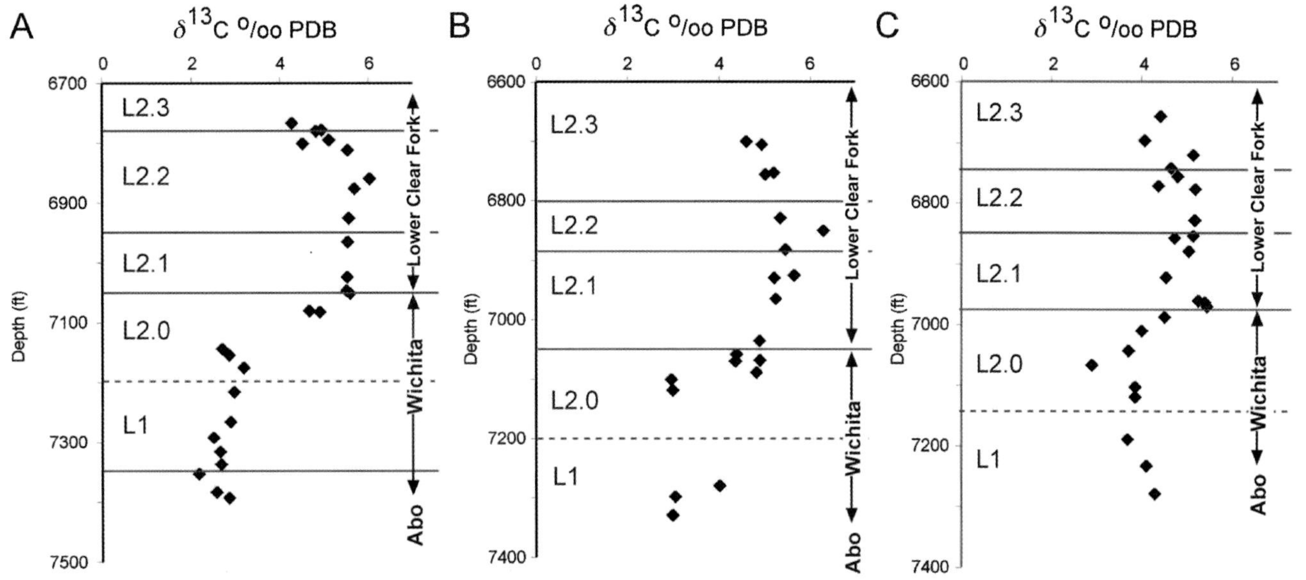

FIGURE 18. Plots of $\delta^{13}C$ with depth for three cores in Fullerton field. Note the close agreement in $\delta^{13}C$ among all three cores, which are arranged along depositional dip from more landward (A) to platform margin (C). (A) FCU C7; (B) FCU 6122; (C) IV-25. Well locations are shown in Figure 4. FCU = Fullerton Clear Fork Unit; PDB = Peedee belemnite.

are present associated with many peritidal sections throughout the reservoir section, the thickest successions are found in the middle of the Wichita section (Figure 7) and at the Wichita-Abo contact in downdip areas of the field. These rocks are variable in fabric but include four basic types: (1) polymict conglomerates, (2) monomict breccias, (3) fractured and tilted beds, and (4) void-filling cement. Although interpreted by some previous workers as mass gravity debris flows, several features make it probable that these breccias were formed by karst-related dissolution.

Polymict conglomerates typically consist of rounded clasts of multiple peritidal and subtidal lithologies (Figure 19). Clasts range in size from a few millimeters to several centimeters in maximum dimension and are commonly subsequent and rounded. Clasts are generally enclosed in mudstone or abut one another at stylolitic contacts. Polymict fabrics (Figure 19A, B) are most common in the middle of the Wichita, with intervals of at least 25 ft (8 m) to as much as 60 ft (18 m) thick encountered in one core (Figure 7). The mixed facies character of these clasts and their roundness indicate that they were formed by sediment transport. However, their association with thick successions of peritidal deposits and their occurrence at a major sequence boundary suggest that a mass gravity transport mechanism is not likely. Instead, their laterally discontinuous nature and their association with other features indicative of karst processes (e.g., Loucks, 1999; Loucks and Mescher, 2001) suggest that they originated as cave-fill deposits.

Related deposits that are also interpreted as the product of karst processes are present at the contact of the Wichita and the Abo (Figures 19C, 20A) at the downdip edge of the field area (FM-1 well, Figure 15). These rocks are commonly characterized by a mixture of two or more facies types, including Wichita tidal-flat facies, green silty carbonate, dark-gray silty carbonate, and Abo subtidal facies (Figures 19C, 20B–E). Instead of being interbedded with one another, the first three appear to overlie the Abo facies in many cores. This relationship suggests that these polymict conglomerates actually represent dissolution and/or erosion of the top of the Abo and subsequent infilling of the irregular surface or differential compaction of the Abo (L1) and overlying transgressive Wichita sediments (L2) at the Abo-Wichita contact. All of these features have been observed in analogous outcrops of the Abo-Wichita contact in Apache Canyon in the Sierra Diablo (Kerans et al., 2000). The outcrop succession reveals that karst features (sinkholes and caves) were formed during the post-L1 sea level fall and then filled with transgressive L2 tidal-flat deposits (Wichita facies). These tidal-flat facies in some instances were brecciated and intermixed with the underlying Abo as sinkholes and collapsed caves.

Monomict conglomerates or breccias consist of broken and rotated clasts of relatively uniform lithology and facies. These rocks are restricted to the Wichita, generally in the middle of the section (Figure 7). Commonly associated with these deposits are fractures, sediment infill (cracks and fissures), void-filling cement

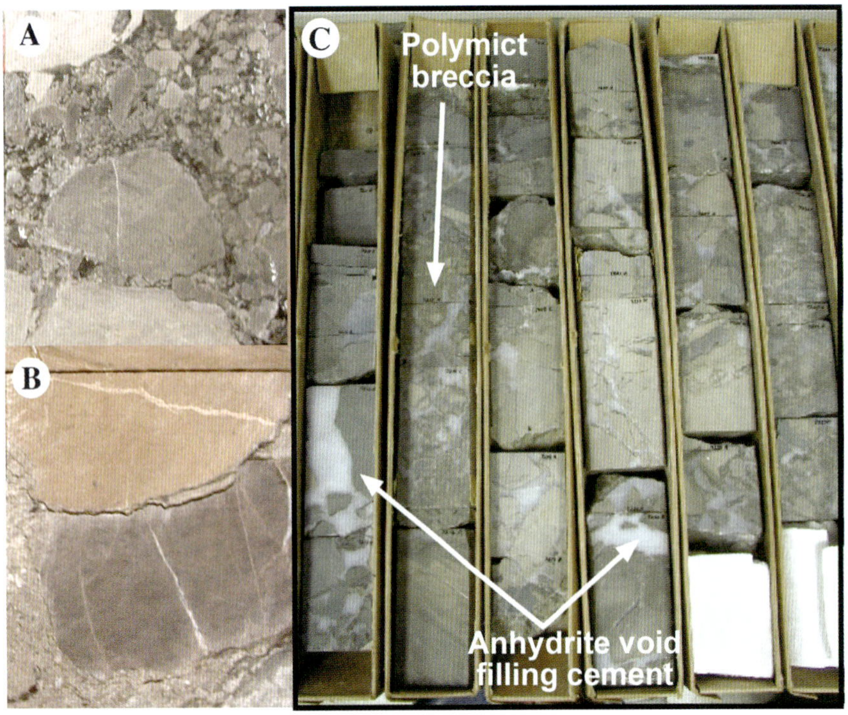

FIGURE 19. Core-slab photographs of breccias and conglomerates of probable karst origin in Wichita Formation. (A, B) Polymict conglomerate with fine-grained breccia clasts filling interclast spaces. (A) FCU 6122; depth, 7201 ft (2195 m). (B) FCU 6122; depth, 7254 ft (2211 m). (C) Breccia succession at Abo-Wichita Amoco University Consolidated IV-25, depths: 7211–7228 ft (2198–2203 m). All cores are 10 cm (4 in.) wide. FCU = Fullerton Clear Fork Unit.

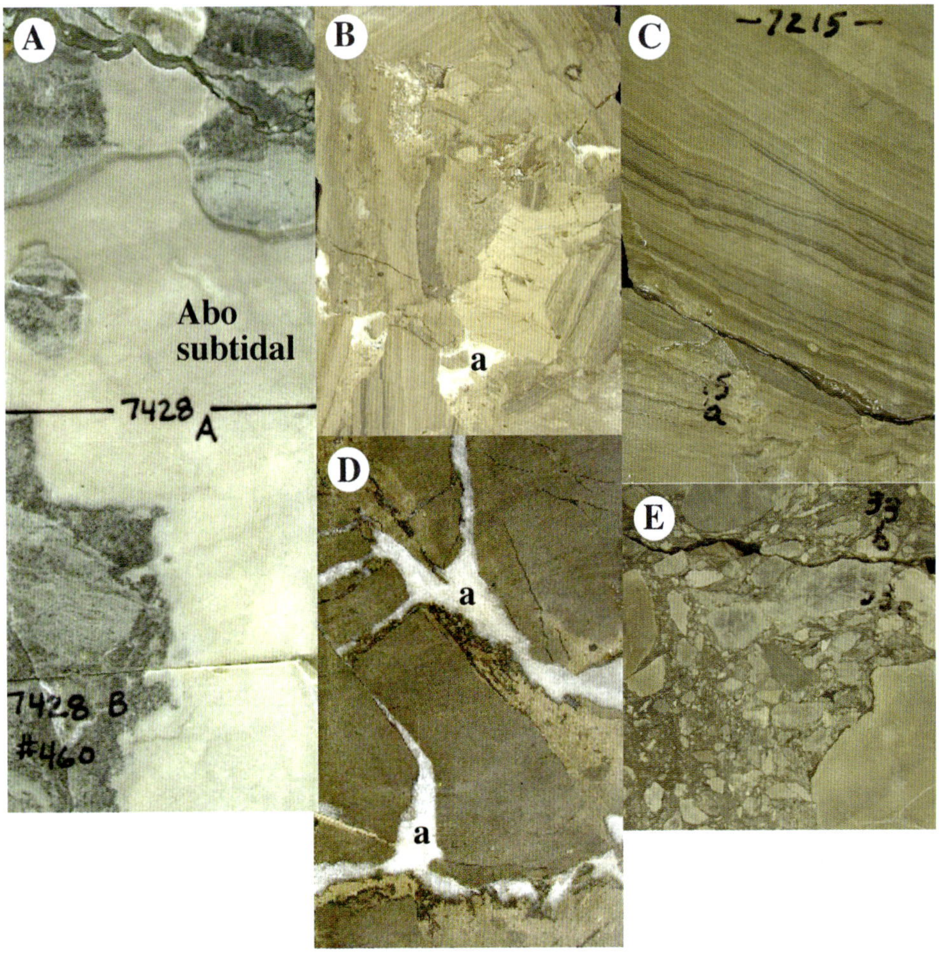

FIGURE 20. Core-slab photographs of karst-related features in the Wichita and Abo. (A) Irregular karsted skeletal packstone at the top of Abo filled with silty, peritidal wackestones of the Wichita. FM-1, 9427 ft (2873 m). (B–E) Monomict breccias from the Wichita. (B) Rotated clasts of tidal-flat facies. (C) Tilted blocks of laminated peritidal facies. Depth, 7215 ft (2199 m). (D) Brecciated clasts of peloidal wackestone. University Consolidated V 15; depth, 6899 ft (2103 m). (E) Clasts of peritidal mudstone-wackestone. All cores are 10 cm (4 in.) wide. a = anhydrite.

(chiefly anhydrite), and other evidence of dissolution. Similar features can also be formed by nonkarst processes in tidal-flat intervals not exposed to true karst (e.g., tepee formation is commonly accompanied by broken and rotated blocks, cracks and fissures, and cement and sediment infill). Some of these rocks, therefore, may not be the result of true karst processes. Their thickness, abundance, and association with other features of karst formation, however, suggest that many are also karst related.

Fractured and tilted beds are also observed in some cores, especially in the middle of the Wichita section. Tilted beds have clearly been formed by postdepositional collapse. In the FCU 5927 well (Figures 7, 15), they form part of a succession that resembles the classic cave-fill–cave-roof succession described by Loucks (1999). This succession consists of 20 ft (6 m) of polymict cave-fill conglomerate composed of mixed tidal-flat facies overlain by 20 ft (6 m) of fractured and locally tilted but apparently generally in-situ beds of tidal-flat and subtidal facies probably representative of a cave roof.

Large zones of void-filling anhydrite cement are also strong indicators of karst-related dissolution and void filling. Zones of massive anhydrite up to 1.5 ft (0.5 m) thick are present in cores in the Wichita at Fullerton field (e.g., FCU 6122, FCU 5927, Figures 7, 15). Smaller anhydrite-filled voids and fractures are ubiquitous within the Wichita and at the Wichita-Abo contact and also point to late cementation of dissolution voids by diagenetic fluids associated with reflux dolomitization of the succession.

Causes and Timing of Karst

Outcrop studies demonstrate that major karsting of the Leonardian sequence occurred at the sea level fall-rise event that is defined by the L1-L2 sequence boundary (Kerans et al., 2000). Cores in downdip areas of the Fullerton field demonstrate this same relationship, where the top surface of the Abo outer-platform facies succession (L1) is karsted and infilled and overlain by brecciated Wichita tidal-flat facies (L2). In updip areas, however, establishing a spatial and temporal relationship between karst formation and the L1-L2 sequence boundary is more problematic. Most karst features in the Fullerton field area are found in an interval of about 150 ft (~46 m) in the middle of the Wichita section (Figure 7). This interval correlates with the interpreted position of the L1-L2 sequence boundary. It should be noted, however, that karst features are developed both below and above this horizon, indicating that karst-related diagenesis was not limited to the L1 sequence. This suggests that two types of karst-related processes occurred.

Primary karsting and dissolution probably occurred during the post-L1 lowstand. At this time, the exposed top L1 surface (consisting of Abo subtidal sediments downdip and lower Wichita tidal-flat sediments updip) developed local caves, sinkholes, and an irregular topography. During the subsequent L2 sea level rise, transgressive peritidal deposits of the Wichita filled the irregular karsted surface, including sinkholes and caves. With continued sedimentation and compaction, parts of the overlying L2 (upper Wichita tidal flat) succession underwent local brecciation and collapse probably because of stress differences set up over underlying karst features. This scenario is consistent with outcrops (Kerans et al., 2000) and fits the distribution of karst features seen at Fullerton field.

Impact of Karsting on Reservoir Quality

Some karst-related deposits exhibit petrophysical properties (porosity, permeability, and saturation) differing at least locally from those of surrounding undisturbed and unaltered deposits. Polymict conglomerates and anhydrite voids are two obvious examples. However, two factors make quantification of the importance of these differences difficult. First, both wireline and core data indicate that most karst deposits do not record significantly different porosity or permeability from that of surrounding nonkarsted deposits. This is probably because most karst fills are composed of the same facies as nonkarsted intervals. Anhydrite void fills are an obvious exception because they contain no porosity or permeability, although most are relatively small and probably of little impact on reservoir flow.

Second, except where cores are available, the distribution of karst features in the reservoir section is not definable. Efforts to identify karst fill using logs and 3-D seismic appear to be negated by the lithologic and petrophysical properties that these features share with surrounding rocks. It is tempting to conclude from these observations that karst features have no impact of reservoir heterogeneity or fluid flow. However, anomalies in water production and flow rates in the Wichita cannot be readily explained by matrix petrophysical properties (T. Anthony, 2003, personal communication). These phenomena may be the result of karst development and postkarstification diagenetic processes.

RESERVOIR IMAGING

Accurate definition of reservoir architecture and distribution of rock fabrics within this architecture is key to defining improved methods for recovery of hydrocarbons remaining in these systems. We used several methods to better image the reservoir at Fullerton field. Especially important in defining the geologic architecture of the reservoir are (1) calibration and use of borehole image logs to aid in the identification and mapping of facies, cyclicity, and rock fabrics and (2) use of 3-D seismic data to constrain the geologic framework.

Identifying Facies and Cyclicity from Borehole Image Logs

Image logs are underused in characterization of carbonate reservoirs. To most, the principal use of such logs is in identification of fractures. However, image logs also have the potential to accurately image many matrix properties that are key to proper characterization, modeling, and exploitation of carbonate reservoirs (e.g., Hammes, 1997).

Traditionally, cores are required to provide key hard data for constraining the distribution of reservoir facies, cyclicity, and rock fabrics. However, when properly calibrated with core observations, borehole image logs can provide most of the data obtainable from cores. This premise is illustrated by an image log from a well in the field (FCU 2564, Figure 4). Although no core was taken in the well, calibration of images to core facies and properties was accomplished using nearby cored wells.

Examination, correlation, and comparison of the image log with nearby cores showed that seven facies can reliably be identified from the image log. This number is fewer than the 12 defined from core studies but is sufficient to provide information necessary to define major facies successions and cyclicity and to provide a strong basis for accurate correlation and interpretation of facies and cyclicity to nearby wells. Facies recognizable on the image log include (1) tidal-flat facies, (2) peloid wackestone-packstone, (3) nodular wackestone-packstone, (4) cross-bedded grainstone, (5) fusulinid wackestone-packstone, (6) karst breccia, and (7) clay-rich mudstone. Although not uniquely definable from the image log, siltstone-sandstone facies can be identified using other wireline logs.

Tidal-flat facies are principally characterized on image logs by their closely spaced parallel and horizontal laminations (Figure 21A). The common presence of fenestral pores and small burrows in tidal-flat facies is typically shown on image logs by abundant, small, black (low resistivity indicating fluid-containing open pores) spots. Tepee structures and sheet cracks are also well imaged where present. Identification of tidal-flat facies is crucial to define both reservoir architecture and reservoir quality. Because they generally occupy cycle tops, their definition makes it possible to define cycle boundaries and thereby facilitate cycle-scale correlation. Equally important is identification of tidal-flat facies and their petrophysical significance. As discussed previously, tidal-flat facies commonly display relatively high porosities but low permeability. Accurate reservoir-quality mapping is, therefore, critical to distinguish tidal-flat rocks from subtidal rocks that are typically higher in permeability.

Fusulinid wackestones (Figure 21B) are identified on image logs as small black (low resistivity) patches that are, in some cases, similar to those representing fenestral pores in tidal-flat facies. However, fusulinid pores are typically larger and more irregular. Fusulinid facies can also be distinguished from tidal-flat facies by a general lack of closely spaced laminations. Fusulinid-bearing rocks generally represent deepest water facies in the Leonardian succession and are found typically at cycle bases. They are thus are important indicators of sea level rise and guides to cycle definition and correlation.

Although generally rare in the Leonardian, cross-bedded grainstones are well imaged on image logs because of their dipping laminations (Figure 21E). These rocks commonly define subtidal cycle tops and are, thus, guidelines to cycle definition and correlation. In addition, where porous, these grainstones can be very permeable.

Other subtidal rocks are typically difficult to subdivide using image logs. Peloid-skeletal wackestones and packstones commonly display a grainy texture probably caused by their small pore size and intercrystalline porosity (Figure 21C). Burrowing of these rocks is ubiquitous, however, and image logs can readily reveal where this burrowing has caused changes in porosity or promoted differential diagenesis. Nodular wackestones, for example, contain abundant anhydrite nodules that were precipitated in the enhanced permeability pathways created by burrowing organisms. The anhydrite that fills these burrows is easily imaged on the image log as whitish (high-resistivity) subspherical patches. Burrow margins commonly still preserve some of the porosity enhancement caused by burrowing. These high-porosity margins are expressed on image logs as black (low-resistivity) rims around the nodules. Where anhydrite has not filled burrows, the result is patches of high porosity (Figure 21D). These are displayed on the image log as black low-resistivity patches that are commonly misinterpreted as open vugs or molds.

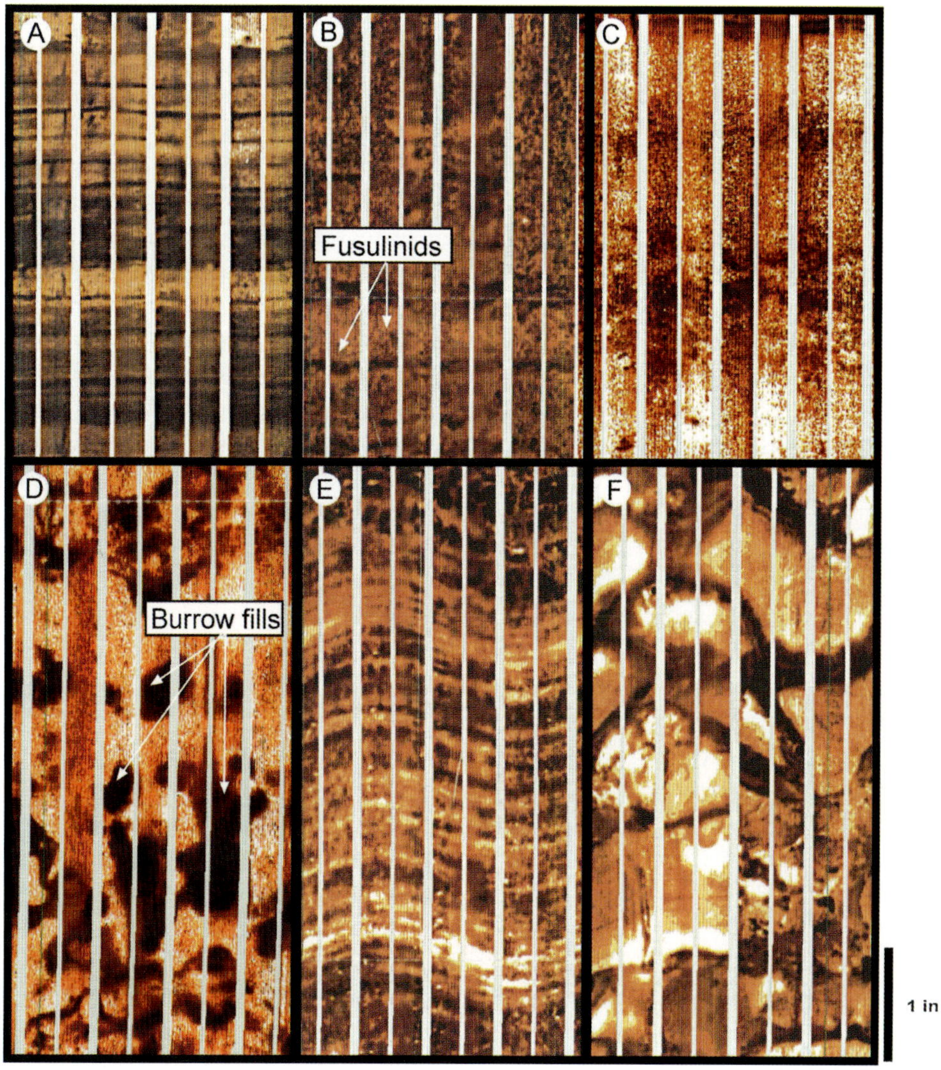

FIGURE 21. Image logs of typical facies and fabrics in the Leonardian. (A) Laminated tidal-flat facies. Depth, 6900–6904 ft (2103–2104 m). Lower Clear Fork, high-frequency sequence (HFS) L2.1. (B) Fusulinid wackestone-packstone facies. Depth, 6990–6994 ft (2131–2132 m). Lower Clear Fork HFS L2.1. (C) Peloid wackestone-packstone facies. Depth, 6905–6909 ft (2105–2106 m). (D) Burrowed wackestone-packstone. Low-resistivity (black) features are commonly interpreted to be open vugs but are actually burrowed zones with higher porosity. Lower Clear Fork. Depth, 6873–6877 ft (2095–2096 m). (E) Cross-bedded grainstone. Lower Clear Fork. Depth, 7002–7005 ft (2134–2135 m). (F) Polymict karst conglomerate. Wichita. Depth, 7171–7182 ft (2186–2189 m). All images from FCU 2564 well. FCU = Fullerton Clear Fork Unit.

Breccias are relatively easily defined by image logs, especially if the clasts are separated by clay-rich or more porous facies that have lower resistivities (Figure 21F). However, the origin of such breccias is not obvious from the images. For example, the image in Figure 21F could represent a tectonic breccia, a deep-water debris flow, or a karst infill breccia. Correct interpretation requires integrated regional and local understanding of depositional setting and diagenetic processes. At Fullerton, as we have discussed, these breccias are clearly the result of karsting.

Other facies, such as siltstone-sandstones and clay-rich carbonate mudstones, can be defined using a combination of image character and wireline-log response. For example, most siltstone-sandstones display parallel horizontal laminations similar to those of tidal-flat facies on image logs and higher than expected values for neutron and acoustic logs. The gamma-ray response is also typically high, although care must be taken to determine whether gamma-ray response comes from potassium-thorium associated with siliciclastics or from uranium, which is less systematically distributed. The spectral gamma-ray log allows this distinction to be made easily. Clay-rich carbonate mudstones (commonly called shales) can be defined by their dark color (indicating low resistivity) and limited thickness. Like siltstone-sandstones, they are also commonly associated with siliciclastics and elevated gamma-ray, neutron, and acoustic-log responses.

Using the calibrated image-log responses just described, we defined a detailed record of facies through an interval of more than 750 ft with a depth range of (6600–7350 ft [2010–2240 m]). Vertical resolution through this interval is potentially better than that attainable from cores (because of uncertainties of core-well depth ties). Cycles defined from these facies successions range in thickness from 3 to 20 ft (1–6 m).

The highest facies and cycle resolution was obtained in intervals characterized by alternating subtidal and tidal-flat facies in the Lower Clear Fork because of the marked contrast in image-log character of these two facies types.

Particularly noteworthy observations from the image log in FCU 2564 is the clear indication that this well occupies a depositional setting significantly different from that of the nearest cored wells to the east. The greater thickness of relatively deep-water fusulinid wackestone-packstone facies in L2.1 here strongly suggests that accommodation increases to the west, which is counter to the regional trend in depositional dip and paleogeography and would not be recognized without this image (or a continuous core). This finding has potential petrophysical significance for the western part of the field by suggesting that deeper water facies of more favorable petrophysical character may be more prevalent in this area than would otherwise be expected. Because of its ability to image key facies and thus facilitate definition of cycles, sequences, and rock fabrics, the image log can contribute critical data that can lead to more accurate reservoir characterization and modeling.

Defining Stratigraphic Architecture and Reservoir Quality from Three-Dimensional Seismic Data

Like image logs, 3-D seismic data are remarkably underused in the characterization of carbonate reservoirs. A critical need in developing approaches that lead to improving recovery from reservoirs containing significant volumes of remaining hydrocarbons is a better understanding of the 3-D distribution of reservoir attributes. Approaches outlined in this chapter for assembling and interpreting well data are the most critical part of this effort. Because they are limited to well control, however, they leave important gaps in our understanding. Three-dimensional seismic data offer valuable data on interwell and extrawell areas (areas of the field where usable well data are absent) that when properly interpreted and applied can greatly improve and, thus, constrain reservoir attribute models. Below are two simple ways in which we used 3-D data to better image both reservoir framework and porosity distribution. A far more detailed and rigorous application of these 3-D data to construction of the geologic model at Fullerton has been described by Zeng (2012).

Constraining Reservoir Architecture

Two-dimensional and 3-D data at Fullerton field provide important guides to stratal architecture of the reservoir succession. Seismic amplitude sections through the entire reservoir interval reveal that much of the section is characterized by generally parallel seismic reflectors (Figure 8). This finding is not unexpected, given the shallow-water platform depositional setting indicated by cores and apparently subhorizontal correlations suggested by wireline logs. However, 3-D and some 2-D data suggest a very different architecture for the basal Leonardian Abo Formation. Seismic data reveal sets of clinoformal reflectors in the Abo that dip generally basinward (toward the east). This apparent clinoformal architecture of the Abo is consistent with observations of clinoformal fusulinid wackestones and packstones in outcropping Abo-equivalent sections in the Sierra Diablo. These clinoforms demonstrate that conventional horizontal correlations of wireline-log data are inappropriate for the Abo. In most cases, resolving cycle-scale correlations of either facies or time surfaces is impossible in such settings.

Defining Reservoir Quality

Seismic data can also be a robust indication of porosity distribution in carbonate reservoirs. Zeng (2012) demonstrated a strong agreement between 3-D seismic impedance data and reservoir porosity at Fullerton field. Because of this robust relationship, even simple amplitude extractions reveal field-scale and fieldwide changes in porosity development that are important to understanding controls on porosity development and for defining interwell and extrawell distribution of porosity.

Figure 22 is an amplitude-extraction map of the HFS L2.1 interval in a small area of the field created to examine possible infill drilling locations. Note that whereas the northern half of the map displays high negative amplitudes indicative of high porosity, the southern half of the area is characterized by markedly lower amplitude. Porosity distribution revealed by this map indicates that only those wells in the northern half of the target area are likely to encounter good porosity in this zone. Note that this information on porosity distribution is obtainable only from the 3-D data volume. The quality of well logs in the area is too poor to determine porosity with any accuracy.

Areas of poor well control like these (caused by either a lack of wells or poor-quality logs) are common through the field, and they greatly compromise efforts to develop a fieldwide strategy to identify and target oil resources. This fact is readily apparent from a comparison of a porosity map based on well logs and a seismic amplitude map of HFS L2.1 (Figure 23). The value of the seismic data volume is especially apparent

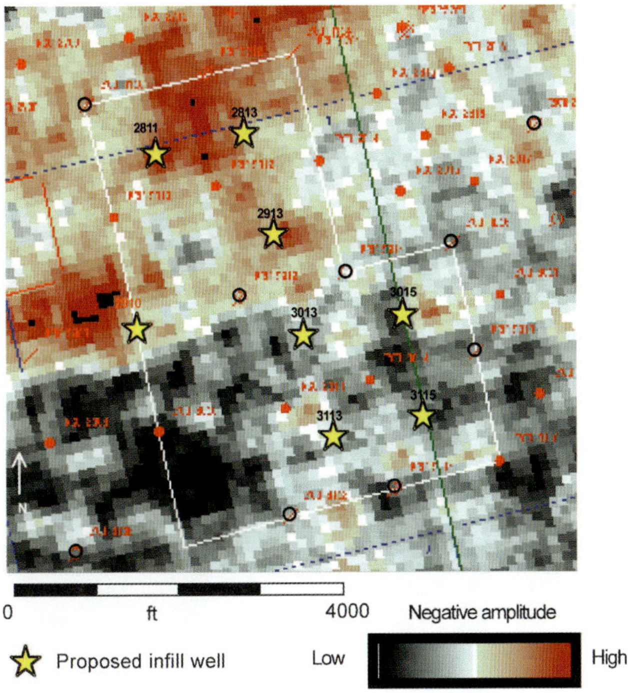

FIGURE 22. **Map of negative amplitude data extracted from Fullerton three-dimensional data. Data show that proposed wells in north half of area will encounter reservoir porosity, whereas those in south half will not. Modified from Ruppel and Jones (2006).**

in the southern end of the field, where limited well control does not accurately image the east–west-trending area of high porosity shown by the amplitude map (cf. Figure 23A, B). Basic amplitude data like these seem to provide a powerful supplement to well control in defining and predicting porosity distribution in the field. Zeng (2012) has documented a more quantitative use of 3-D data in defining porosity development in the field.

RESERVOIR ARCHITECTURE

A critical component of a robust reservoir model is a geologically constrained reservoir framework. Geologically accurate models must be based on correlations of time-stratigraphic units, the most readily correlative of which are cycle and sequence boundaries. At Fullerton field, we used correlations of cycle tops to construct a reservoir model for the Clear Fork. Although these correlations are ultimately based on correlations of wireline logs (porosity and gamma ray), the underlying basis for interpretation and correlation of these logs is the knowledge of one-dimensional and 2-D facies and cycle-stacking relationships developed from integrated studies of cores and outcrops.

Cyclicity and Flow-Unit Definition

Reservoir flow units are most appropriately based on the definition and mapping of depositional cycle boundaries because, when properly defined and correlated, depositional cycle tops represent the best available indicator of original depositional surfaces or timelines. The procedure for identification of cycle boundaries for flow-unit definition was well described by Ruppel and Ariza (2002) and Lucia and Jennings (2002) for the South Wasson Clear Fork reservoir.

Depositional cycle tops must first be identified in cores using techniques described above. These surfaces must then be correlated throughout the field using calibrated well logs. Ideally, correlations should be based on logs that can be tied directly to depositional facies independent of diagenesis or porosity variations. The gamma-ray log in certain ideal settings serves this function. However, throughout most of the Leonardian carbonate succession (in fact, throughout most of the Permian carbonate section in the Permian Basin), the conventional gamma-ray log is not accurate for detailed facies or cycle correlation because of variable volumes of uranium, potassium, and thorium in platform-carbonate successions. The spectral gamma-ray log, however, can help distinguish variations among these elements and, thus, is useful in separating siliciclastic-rich rocks (that contain high levels of potassium and thorium) from siliciclastic-poor rocks. Because small volumes of siliciclastics (mostly clays) are commonly associated with tidal-flat facies in Leonardian rocks, the spectral gamma ray is locally helpful in distinguishing tidal-flat facies from subtidal facies (Ruppel, 1992; 2002). Variations in uranium content are commonly not systematic, however, so total gamma-ray response commonly varies independently of facies. Resistivity logs can also be used locally to separate tidal-flat facies from subtidal facies on the basis of differences in saturation (Ruppel, 2002). However, even under ideal circumstances, neither gamma-ray logs nor resistivity logs can accurately depict facies in Leonardian rocks.

Because of the uncertainties of using gamma-ray logs, we used porosity logs to define facies and cycle tops in the Lower Clear Fork section at Fullerton field. The basis for this approach comes from integrated studies of cores and wireline logs at Fullerton that demonstrate two important attributes of Leonardian cycles. First, cycle-capping facies are either grain-rich subtidal or tidal-flat facies. Second, porosity is most commonly associated with these facies and is, therefore, best developed at cycle tops. These relationships are critical to the use of wireline logs for correlation of

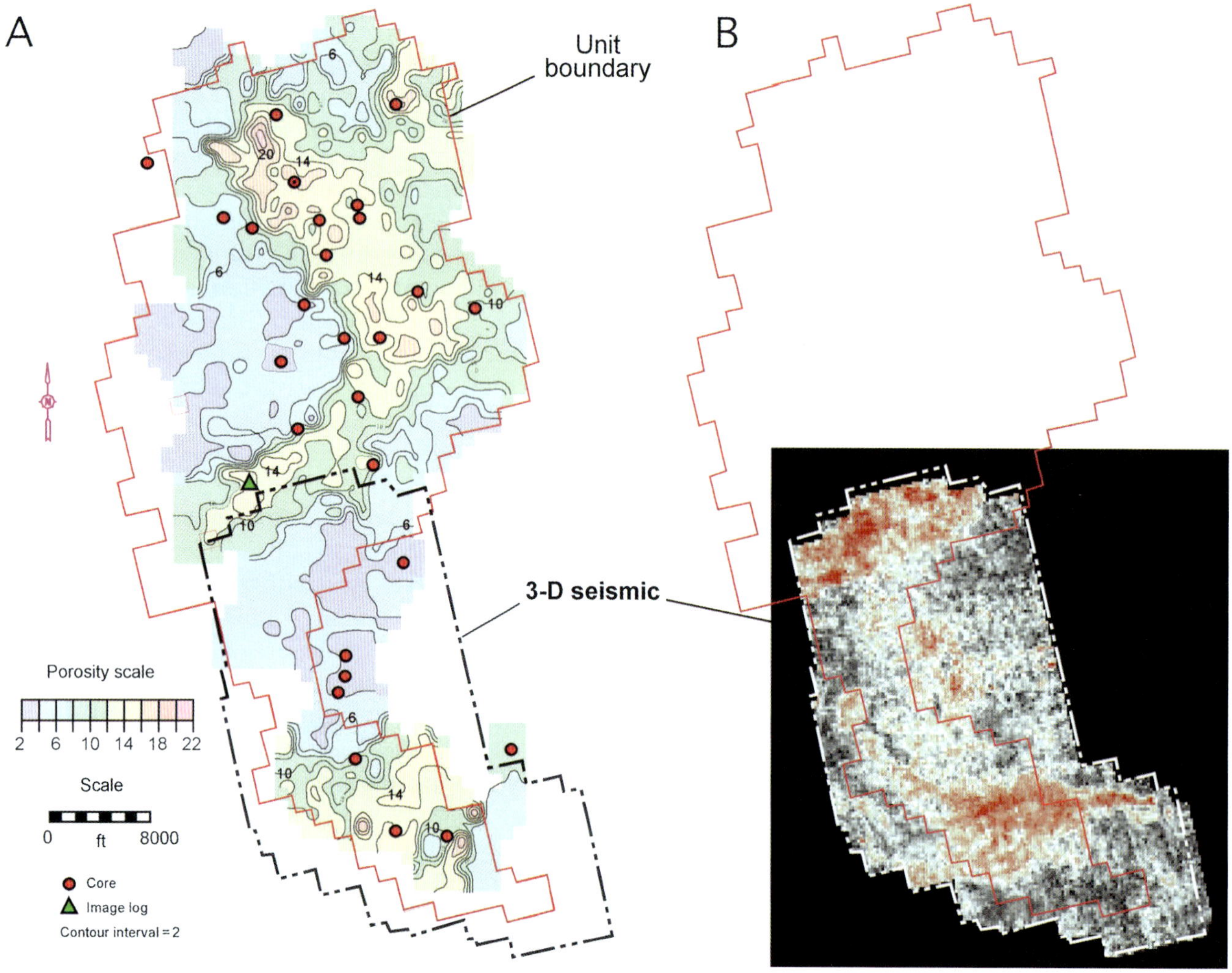

FIGURE 23. Porosity development in high-frequency sequence (HFS) 2.1. (A) Map of porosity derived from well-log calculations. (B) Map of negative amplitude data extracted from Fullerton three-dimensional seismic data. Note that porosity in east–west-trending area of negative amplitude in the southern part of the field is not well imaged by wireline-log data despite close spacing of wells.

cycles and of flow units. If properly constrained by core and outcrop calibration of facies, porosity, and cyclicity, correlation based on porosity response is in many cases the most geologically sound basis for constructing cycle correlations and establishing a basis for true flow-unit correlations.

It is important to establish geologically defined cycle tops at the highest resolution possible. Outcrop studies demonstrate that cycles less than 10 ft (<3 m) thick can be correlatable over large areas of the platform. Where wireline data permit, correlations should be taken to this level of resolution through the reservoir even if later upscaling is planned for reservoir modeling. High-resolution cycle correlations are important for two reasons. First, coarser scale correlations are much more likely to be in error. At Fullerton, we found that early coarser scale correlations made at the sequence or cycle-set scale were later proven to be in error by a cycle or two after we recorrelated the succession at the higher resolution cycle scale. From a reservoir modeling point of view, therefore, flow units defined by coarse-scale correlations are more likely to cross-connect flow layers than those defined by finer scale correlations. Second, upscaling (grouping of cycles into thicker flow-unit packages for modeling) is more likely to retain original geologic architecture if based on fine-scale (cycle-scale) correlations. Accordingly, we attempted to correlate the reservoir succession at the highest possible level of detail, supported by outcrop and core observations and wireline resolution.

Lower Clear Fork Reservoir Cycle Architecture

The robustness of the use of porosity logs for defining facies and cyclicity is apparent from core and

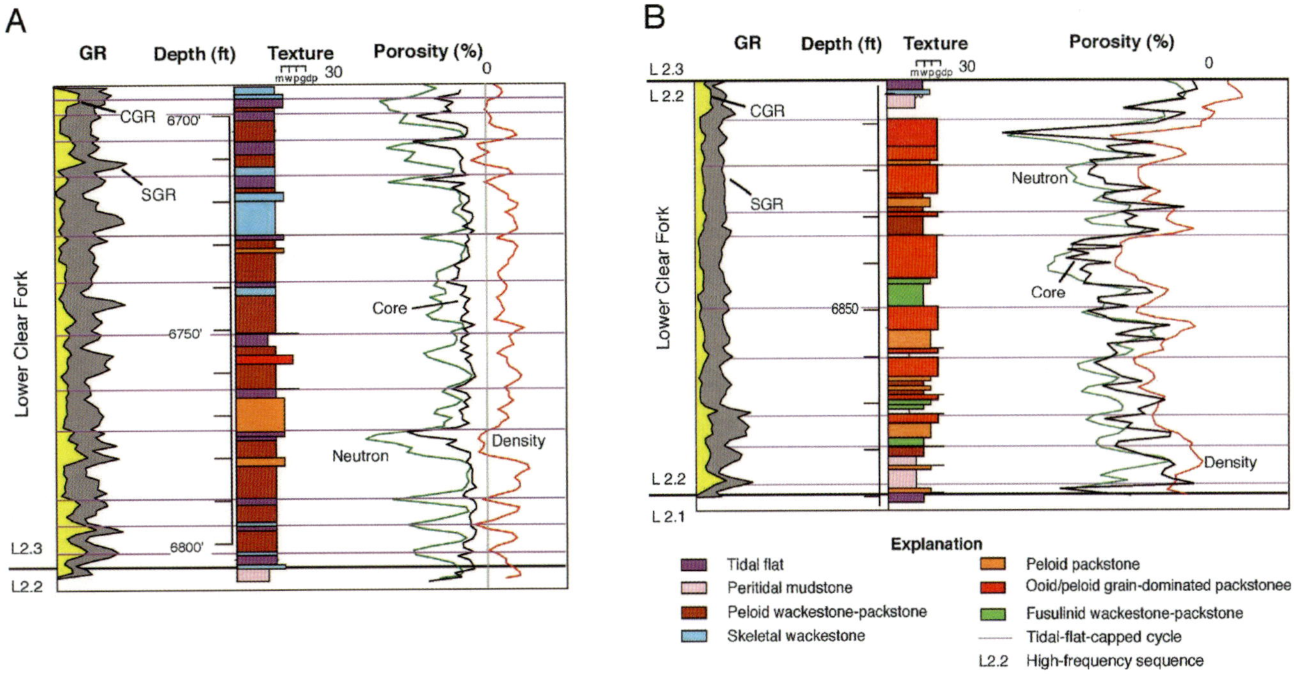

FIGURE 24. Comparison of core-defined facies and cyclicity with porosity logs in the Lower Clear Fork. (A) Tidal-flat–capped cycles of HFS L2.3. (B) Subtidal cycles of HFS L2.2. FCU 6122. HFS = high-frequency sequence; FCU = Fullerton Clear Fork Unit; m = mudstone; w = wackestone; p = packstone; gdp = grain-dominated packstone; GR = gamma-ray log; SGR = spectral gamma-ray log; CGR = computed gamma-ray log. Modified from Ruppel and Jones (2006).

log relationships in HFS L2.3. A comparison of core and log porosity shows that porosity is nearly entirely associated with cycle-top tidal-flat facies (Figure 24A). Porosity logs can, thus, be used to define both facies and cyclicity. Although not an oil-producing part of the reservoir, the L2.3 section demonstrates the porosity-log calibration that underlies all cycle-based correlations in the field. It also underscores the importance of examining even nonproductive parts of the reservoir succession during reservoir characterization activities. The L2.3 cycles defined from porosity-log response average about 10 ft (~3 m) in thickness and are relatively easily correlated across the field.

The same approach to log-based cycle correlation is applicable to subtidal sections in the Lower Clear Fork. However, in these cases, cycle tops are typically composed of grain-rich subtidal facies. Here too, however, a strong relationship in cored wells exists between cycle-top facies and porosity (Figure 24B).

Using porosity logs, we defined and correlated 15 cycles in the Lower Clear Fork reservoir section at Fullerton (L2.1, L2.2; Figure 3). The average thickness of these cycles is about 17 ft (~5 m) for the entire section of L2.1 and L2.2. However, cycle thickness varies systematically by sequence within this interval. The HFS L2.2 cycles average about 11 ft (~3 m) in thickness, whereas the L2.1 cycles, with the exception of low-accommodation tidal-flat cycles at the base and top of the sequence, are nearly twice as thick (Figure 3). This difference is probably caused by two factors. First, L2.1 deposits record maximum flooding of the platform and probably development of maximum accommodation. Overall upward thinning of cycles from L2.1 to L2.2 is consistent with an overall upward decrease in accommodation. Second, lateral changes in facies stacking in L2.1 cycles suggest local variations in sediment accumulation patterns that may have been caused by topographic relief on the platform during L2.1 flooding.

Wichita Reservoir Cycle Architecture

For the Wichita, it is not possible to rigorously define cycle boundaries because of the preponderance of low-accommodation tidal-flat facies. Both core and outcrop studies show that rocks deposited in such settings rarely display systematic trends in vertical facies stacking and generally exhibit low lateral facies continuity. Accordingly, patterns of depositional facies do not define extrinsic controls (for example, sea level rise and fall) but, instead, local controls on sediment accumulation (e.g., paleotopography and climate).

In part of the Wichita section, however, it is possible to define cyclicity based on mineralogy and porosity. In the northern part of the field, the upper

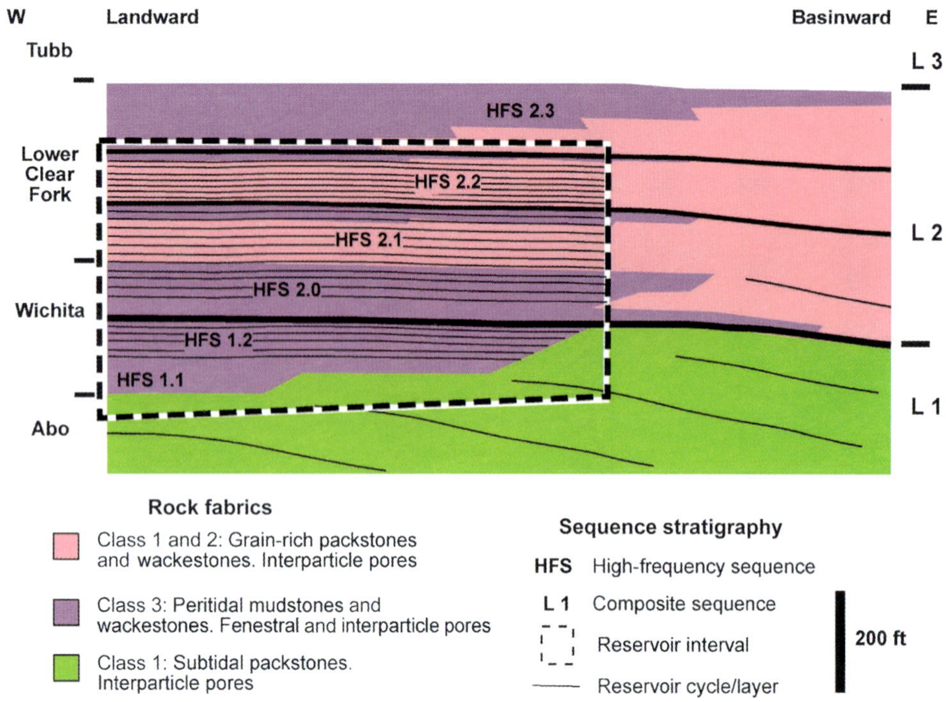

FIGURE 25. Schematic depiction of the reservoir framework used for model construction at Fullerton Clear Fork reservoir showing interrelationships of stratigraphic sequences, rock fabrics, and cycle-based reservoir framework. HFS = high-frequency sequence. Modified from Ruppel and Jones (2006).

Wichita contains multiple intervals of low-porosity limestone interbedded with higher porosity dolostone (Figure 7). Because these dolostone-limestone cycles are generally parallel to overlying and underlying cycle and sequence boundaries and because they are relatively continuous (Figure 7), we believe that their formation was related to cycle deposition, exposure, and diagenesis. Early dolomitization of peritidal cycle tops has been widely documented in low-accommodation carbonate-platform successions of all ages. In this model, limestone beds represent undolomitized cycle bases, whereas intervening dolostone beds represent early dolomitized cycle tops. Oxygen isotope values for the limestone beds are typical of calcite precipitated from middle Permian seawater and, thus, are consistent with such a model. This implies that the architecture of these diagenetic cycles is closely tied to depositional surfaces and thus represents a good approximation of time surfaces. These surfaces are also especially useful from a reservoir point of view because they define layers of low and high porosity and permeability. The top four cycles in the Wichita are defined on the basis of these cyclic dolostone (high-porosity)-limestone (low-porosity) couplets (Figure 3).

Where limestone is not present in the Wichita (southern part of the field and lower Wichita throughout the field) (Figure 7) (Harrington and Lucia, 2012), this approach is not possible. In these areas, porosity alone must be used to construct the reservoir framework. Accordingly, the framework established for these parts of the Wichita is far less geologically robust than that defined for other parts of the reservoir. For this part of the reservoir, we defined and correlated 12 surfaces on the basis of porosity. We are confident that these surfaces are subparallel to time surfaces, given their parallelism to overlying Lower Clear Fork cycle-top surfaces and to the middle Wichita marine floodback that marks the L1-L2 boundary. However, we cannot tie them rigorously to cyclicity.

Abo Reservoir Architecture

The fact that the Abo is dominated by clinoformal bedding typical of outer-ramp carbonate deposits is well demonstrated by both outcrop studies (Kerans et al., 2000) and 3-D seismic data at Fullerton (Figure 8). Abo architecture, therefore, differs significantly from the generally subparallel, horizontal architecture of the platform-top Wichita and Lower Clear Fork successions. Outcrop studies and seismic studies at Fullerton and elsewhere (Kerans et al., 2000; Ruppel et al., 2000) also demonstrate that clinoformal outer-platform successions like the Abo do not contain readily correlatable cyclic successions because of multiple and changing sources and distributional patterns and extreme variations in lateral textures and fabrics of these transported sediments. Accordingly, it is not possible to establish an accurate internal architecture for these deposits using wireline logs or 3-D seismic. For purposes of reservoir modeling, we created a series of conceptual clinoform surfaces to constrain the model architecture. Although these surfaces do not accurately describe the architecture of the Abo, they do illustrate the nonparallel and nonhorizontal nature of these deposits.

Reservoir Model

Details of reservoir models developed for the Clear Fork reservoir at Fullerton are discussed in detail by Wang and Lucia (2012). However, the basic architecture of the reservoir framework used in the models is depicted in Figure 25. Key aspects of this model are (1) subhorizontal and subparallel nature of Lower Clear Fork and upper Wichita (L 2) surfaces, (2) effective pinchout of lower Wichita (proximal L1) surfaces at the facies change from Wichita tidal flats to Abo subtidal, and (3) clinoformal nature of Abo (distal L1).

Patterns of Porosity Distribution

Core and log data from wells that penetrate the Abo indicate that the Abo locally contains high porosity. However, well control is too sparse and incomplete to map porosity distribution accurately. Wang and Lucia (2012) give an estimation of the 3-D distribution of Abo porosity using a full-field model based on available porosity data and a conceptual geologic framework. However, the accuracy of this distribution must be considered relatively low because of the scarcity of Abo data.

The considerable well-log data for the Wichita Formation demonstrate that it contains higher porosity × height (phi × h) than either the Abo or the Lower Clear Fork. The highest total phi × h lies along the margins of the field structural crest (cf. Figure 26 with Figure 4), suggesting that porosity development may have been a function of structure. However, when phi × h is examined by stratigraphic horizon, this interpretation seems less obvious. The key element of porosity distribution in the lower parts of the Wichita is the marked decrease in phi × h to the southeast because of the southeastward facies transition from Wichita tidal flat to Abo subtidal facies (and accompanying decrease of Wichita thickness) along the margin of the inner platform (Figure 27).

The lower part of the upper Wichita (Figure 28), which represents the base of the L2 composite sequence and the beginning of L2 transgression (HFS L2.0), displays a well-defined arcuate trend of porosity that generally parallels the underlying L1 Wichita-Abo facies transition (Figure 27). Note that both wireline-log data (Figure 28B) and 3-D seismic amplitude extraction data (Figure 28C) image this trend well. The

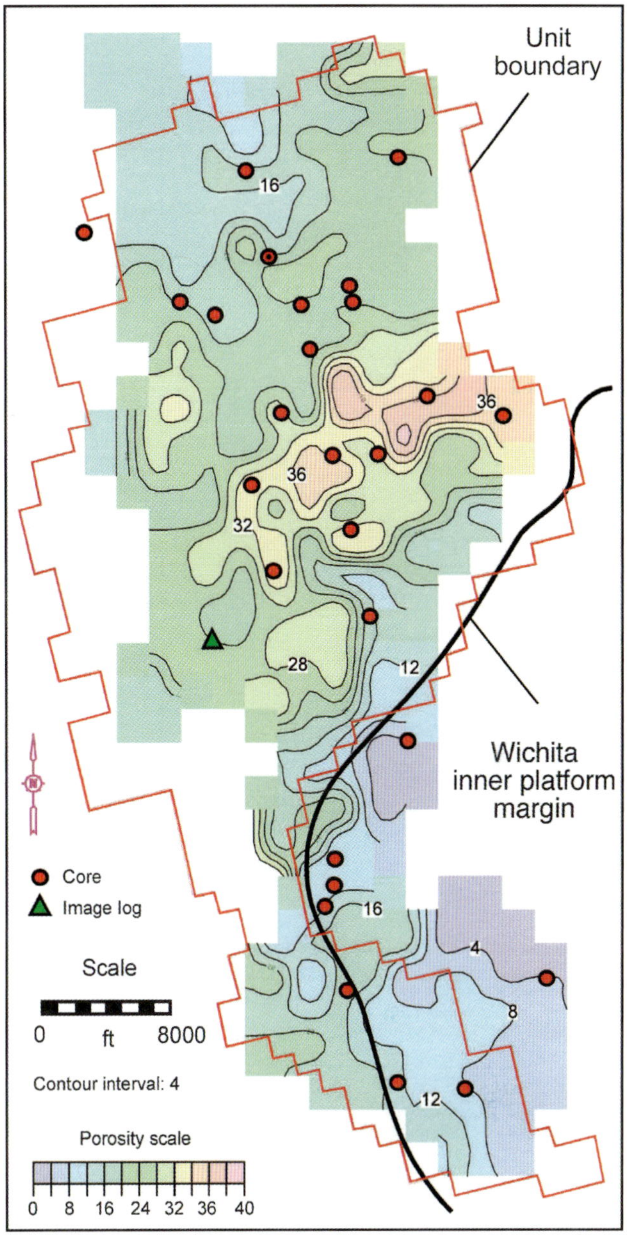

FIGURE 26. Map of total porosity × height in Wichita. The low values are in the southeast corner of the field mostly because of the change of facies from Wichita tidal-flat facies to Abo facies in that area.

transition from L2.0 upper Wichita rocks to L2.0 Lower Clear Fork rocks is located just east of this belt of porosity. The decrease in porosity southeast of this belt is, thus, associated with a change from tidal-flat rocks to subtidal-flat rocks—presumably the result of changes in bathymetry and accommodation downdip on the lower Wichita-Abo platform. The decrease in porosity to the northwest, however, is not associated with any apparent change in depositional facies. Accordingly, the cause of porosity development must be more a

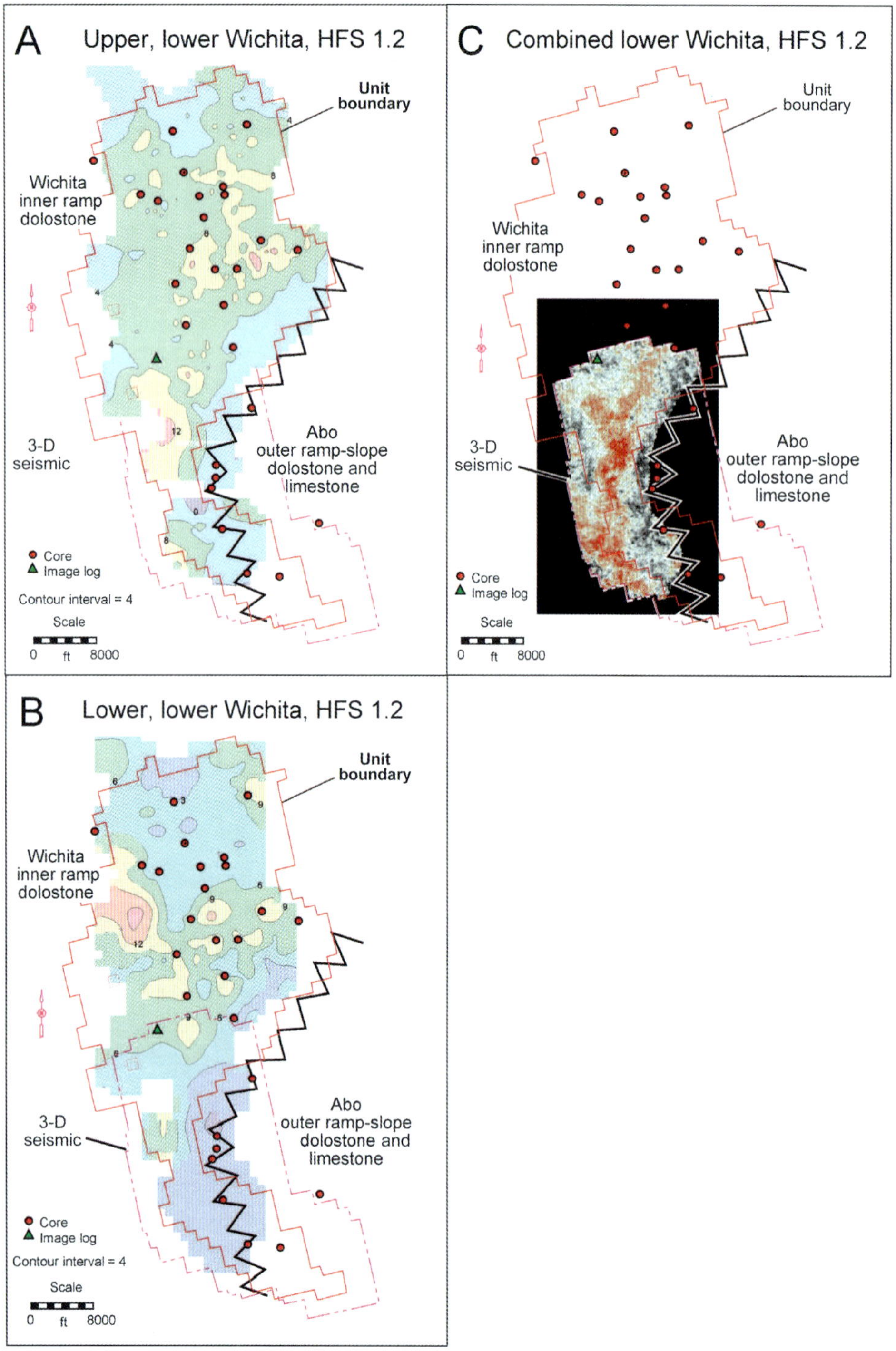

FIGURE 27. Maps of porosity distribution in the lower Wichita (sequence L1). (A) Porosity × height (phi × h) map from wireline logs in the upper part of the lower Wichita. (B) Phi × h map from wireline logs in the base of the lower Wichita. (C) Amplitude extraction from three-dimensional seismic volume for the combined lower Wichita section. High negative amplitude equals high porosity. The zigzag line represents Wichita-Abo facies transition. Lower Wichita rocks are absent at the east of this line. HFS = high-frequency sequence.

function of diagenesis than deposition. A model of this diagenesis is presented in the following section.

The uppermost Wichita also displays strong differential development of porosity across the field that does not appear to be related to depositional facies. Areas of highest porosity are located in the north and south parts of the field. The low porosity is encountered in the center and northwestern parts of the area (Figure 28A). Like the lower Wichita, systematic differences between tidal-flat deposits are not apparent across the field. These variations must, therefore, be the result of differential diagenesis. Porosity patterns suggest that the upper Wichita porosity may exhibit a similar, albeit westward-displaced, arcuate trend much like that in the lower Wichita. The high-porosity area in the north extends southwestward and may continue west of the field to link up with porosity in the southern end of the field. If so, it would imply a link

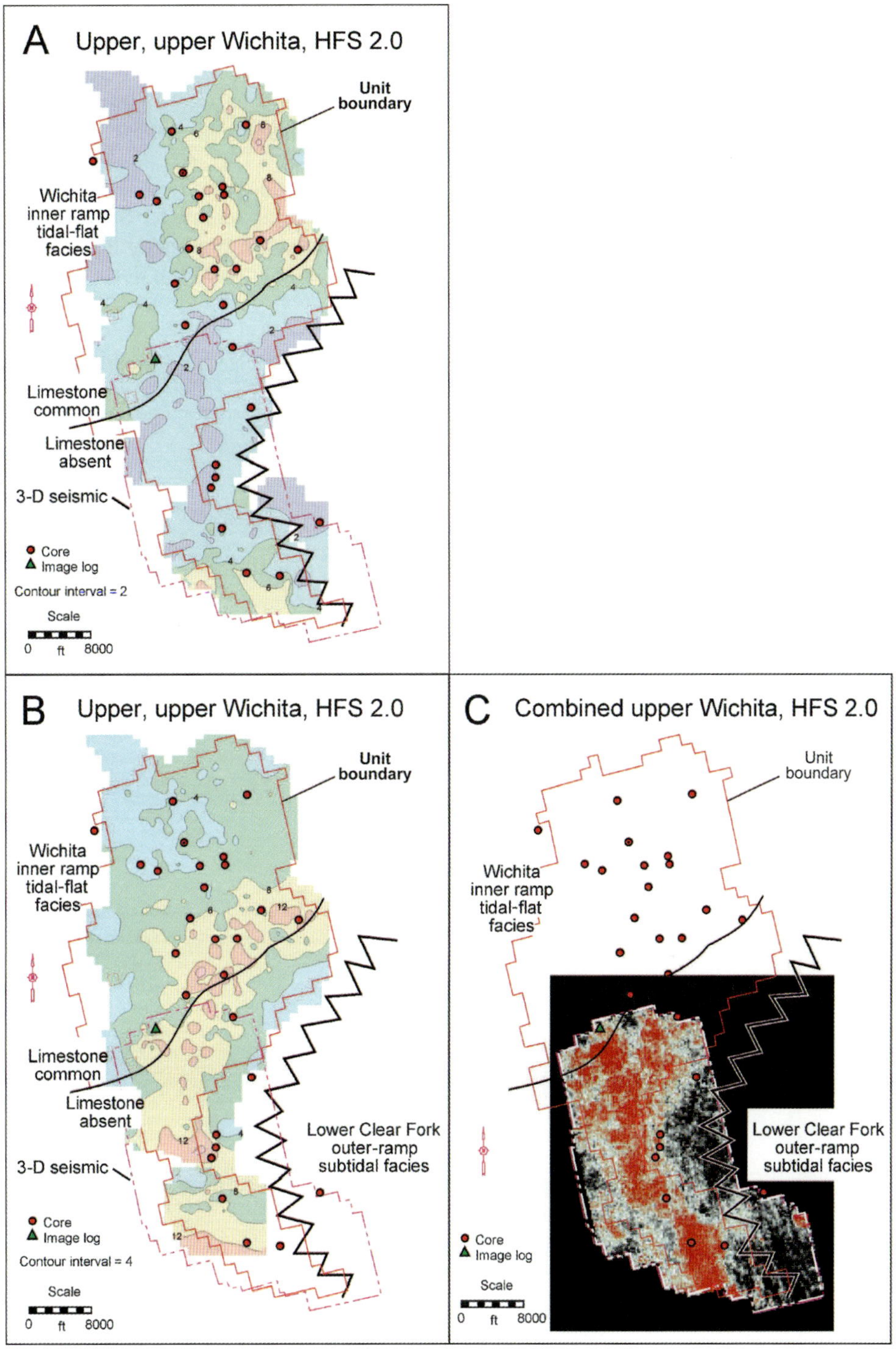

FIGURE 28. Maps of porosity distribution in the upper Wichita (HFS L2.0). (A) Porosity × height (phi × h) map from wireline logs in the upper part of the upper Wichita. (B) Phi × h map from wireline logs in the lower part of the upper Wichita. (C) Amplitude extraction from three-dimensional seismic volume for the combined upper Wichita section. High negative amplitude equals high porosity. The zigzag line represents Wichita-Lower Clear Fork facies transition. Upper Wichita rocks are absent at the east of this line. Note that the porosity concentrated along an arcuate belt following the outer edge of the Wichita inner-platform tidal-flat complex. HFS = high-frequency sequence.

between reservoir development and platform geometry similar to that suggested for the lower Wichita. The more westward development of this porosity trend is consistent with progressive transgression of the L2 platform and backstepping of the Wichita tidal-flat systems tract. Limestone is common in the upper Wichita tidal-flat succession throughout the northern part of the field but is unrelated to areal trends in porosity. These nearly impermeable limestones do, however, affect fluid flow through this part of the reservoir.

The Lower Clear Fork HFS L2.1 displays an arcuate trend in porosity development similar to that seen in the upper Wichita (see map of wireline porosity in L2.1, Figure 29B, C). Like the uppermost Wichita (Figure 28A), a strong suggestion exists that the northern trend may extend west of the field and curve south to link up with the southern areas of porosity, as does

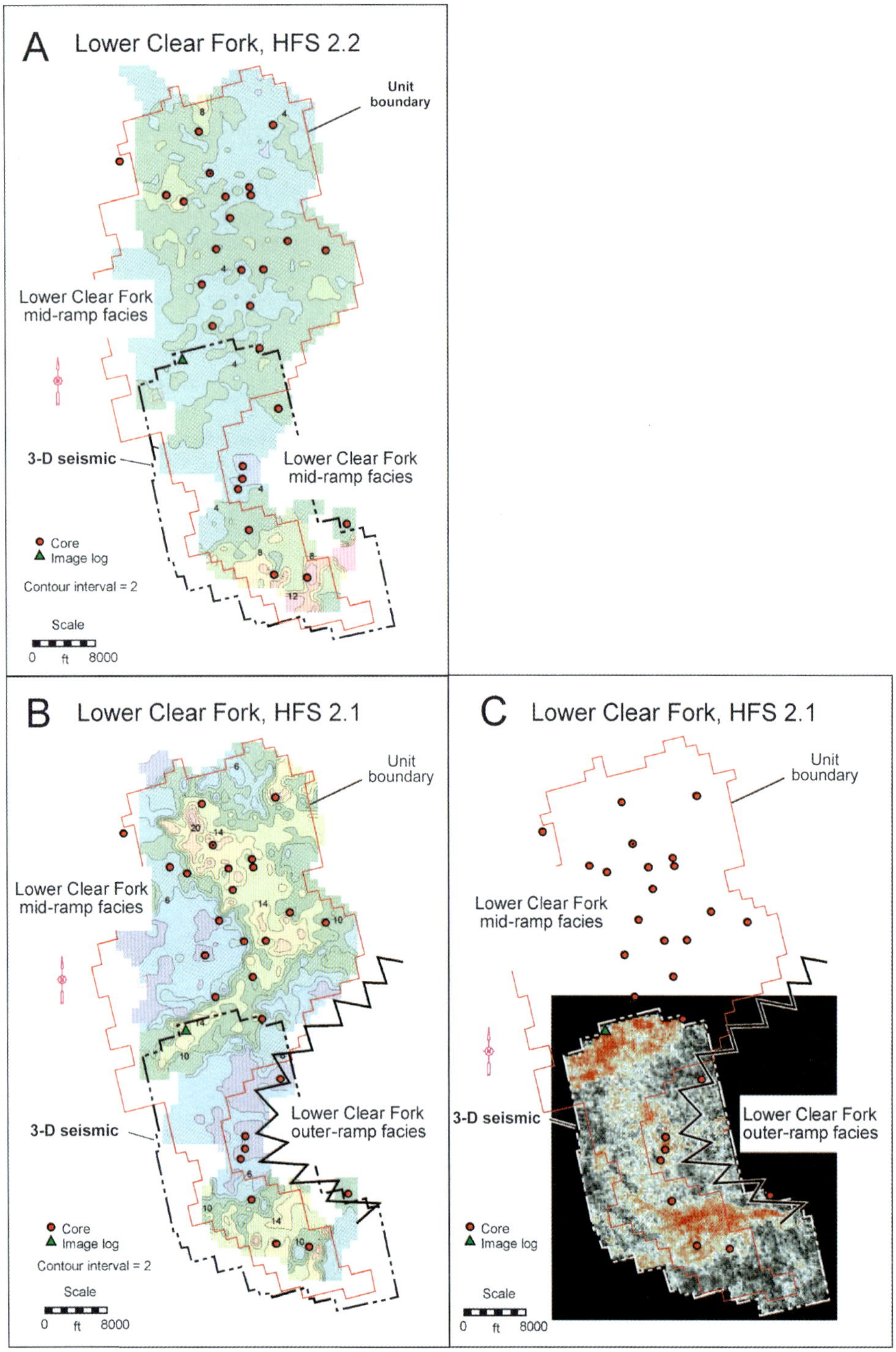

FIGURE 29. Maps of porosity distribution in the Lower Clear Fork. (A) Porosity × height (phi × h) map of HFS 2.2 from wireline logs. (B) Phi × h map of HFS L2.1. (C) Amplitude extraction from three-dimensional seismic volume for the HFS L2.1 interval. High negative amplitude equals high porosity. The areas of high porosity in HFS L2.1 and HFS L2.2 correlate with the areas of abundant limestone (cf. Figures 23, 25). HFS = high-frequency sequence.

the lower upper Wichita (Figure 28B). As is the case with Wichita porosity development, depositional facies do not appear to be the controlling factor in the development of this trend. The eastward decrease in porosity (in the southern part of the field) is associated with a gradual change from middle-ramp grain-rich packstones to outer-ramp fusulinid-rich wackestones. The area of high porosity may represent a low-energy ramp crest because the abundance of ooid-skeletal grain-dominated packstones and grainstones is somewhat greater. But the change in depositional facies associated with the major decrease in porosity into the northwestern part of the field is subtle (Figure 29B). Instead, like the Wichita, the high porosity in L2.1 is most likely the result of dominantly diagenetic patterns. A direct relationship exists between limestone abundance and high porosity in these rocks. Areas of high porosity in the northern part of the field contain

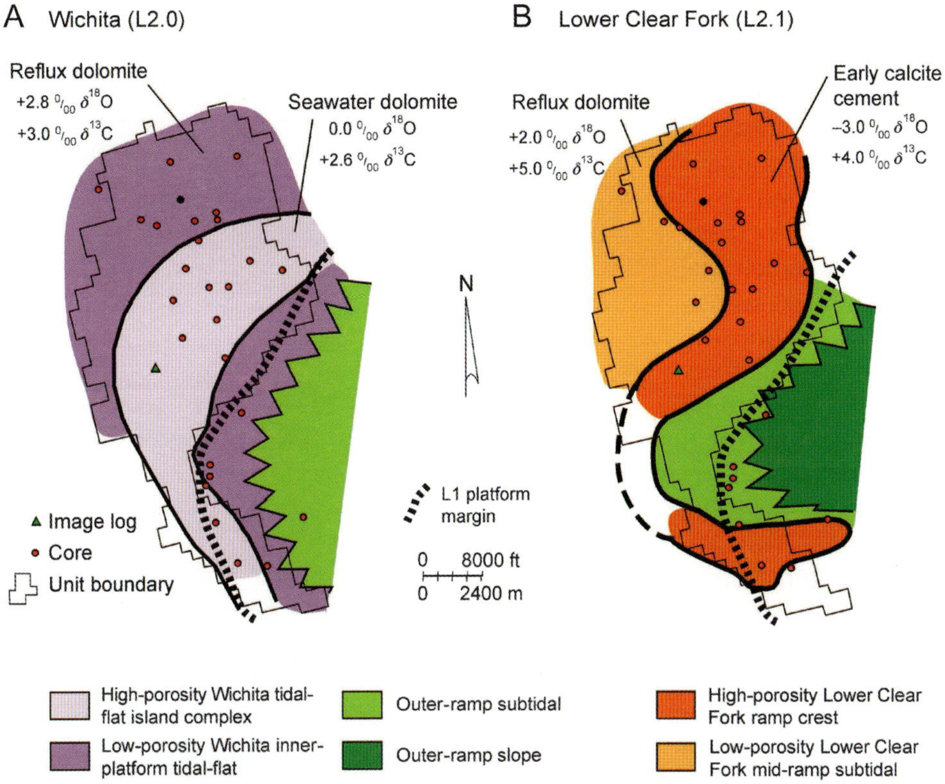

FIGURE 30. Distribution of facies, porosity, and stable isotope data in the upper Wichita and the Lower Clear Fork (L2 sequence). (A) Upper Wichita (HFS L2.0). Note that porosity is highest in tidal-flat dolostones that display seawater dolomite $\delta^{18}O$ signatures. Lower porosity dolostones display enriched $\delta^{18}O$ values indicative of diagenesis by concentrated reflux brines. (B) Lower Clear Fork (HFS L2.1). Note that porosity is highest in grain-rich ramp-crest limestones that display seawater calcite $\delta^{18}O$ values. Lower porosity dolostones exhibit enriched $\delta^{18}O$ values that, like the underlying Wichita, suggest diagenesis by concentrated reflux brines. HFS = high-frequency sequence. Modified from Ruppel and Jones (2006).

mixtures of porous dolostones and porous limestone; the high-porosity area at the southern end of the field is nearly all limestone.

No obvious trend in porosity is apparent in HFS L2.2 (Figure 29A). However, like L2.1, areas of highest porosity are associated with limestone. For L2.2, this area is dominantly in the southern end of the field (see Harrington and Lucia, 2012).

Models for Porosity Development

Although porosity is developed in all parts of the Abo-Wichita-Lower Clear Fork reservoir section, obvious areal variations in porosity exist that reflect a combination of depositional, diagenetic, and structural controls on reservoir development. Most of the major reservoir intervals exhibit spatial variations in porosity development that parallel platform paleotopography, particularly in the lower and upper parts of the upper (L2) Wichita and in the Lower Clear Fork (HFS 2.1) (Figures 28, 29). In each of these cases, trends of high porosity are situated immediately up the depositional dip from the position of the underlying Wichita-Abo facies boundary. However, the mechanism for porosity development must differ between the Wichita (Figure 28) and the Lower Clear Fork (Figure 29).

In the case of both Wichita intervals, the depositional facies in the high-porosity trend are essentially identical with those in low-porosity areas both depositionally updip and downdip. This uniformity of facies indicates that porosity formation is the result of local diagenetic processes, instead of variations in original depositional texture. Stable isotope data collected in this study combined with those reported by Ruppel and Jones (2006) suggest that high-porosity dolomites in the Wichita were formed by seawater-dominated fluids, perhaps mediated by meteoric mixing. This is indicated by light $\delta^{18}O$ values (average, +0.0‰ PDB) for these dolomites (Figure 30A). Restriction of high-porosity rocks to the downdip margin of the Wichita tidal flat suggests that porosity development may be the result of early seawater-dominated dolomitization and stabilization along the seaward margin of the Wichita tidal flat. Available data suggest that less porous, updip Wichita, tidal-flat deposits were dolomitized during a later event by very different fluids. These rocks contain heavier isotopes ($\delta^{18}O$ = 2.78‰ PDB; Figure 30A) that are typical of younger Lower and Upper Clear Fork rocks and are more indicative of brines that were evaporatively concentrated (Saller and Henderson, 1998; Ruppel, 2002). Possible sources of such brines, given stratigraphic data, are updip tidal flats developed at the tops of sequences L2.1, L2.2, and L2.3 or those developed during L3 transgression (Tubb deposition). The presence of limestone interbeds in the Wichita in the northern less porous part of the field suggests such a distal diagenetic setting (one that is distal to the source of diagenetic fluids). The lack of porosity in the updip Wichita tidal flats may reflect compaction and porosity reduction caused by dolomitization and sulfate emplacement during Clear Fork–Tubb reflux dolomitization.

The Lower Clear Fork HFS L2.1 also displays a trend in porosity development that closely parallels the Wichita-Abo and Wichita-Clear Fork facies transitions (Figure 29). In this case, however, a relationship exists between both facies and mineralogy and porosity development. Facies data suggest that the trend of high porosity may represent a platform-ramp crest, an area characterized by more common ooid-bearing grain-rich packstones and grainstones and by locally more abundant tidal-flat caps. Ramp-crest development was probably controlled by inherited paleotopography over the L1 (Wichita-Abo) inner-ramp margin and/or differential subsidence over deep-seated faults in the same way that the position of the Wichita-Abo facies transition is most likely controlled by such deep structures. Mineralogical data show that this trend is also an area of abundant calcite; lower porosity areas to the east and west are nearly entirely dolomite (see Harrington and Lucia, 2012). As previously discussed, the oxygen isotope data from these calcites ($\delta^{18}O$ = –3.0‰) indicate that they are essentially unaltered original marine precipitates (Figure 30B). Updip dolomites, by contrast, display $\delta^{18}O$ values ($\delta^{18}O$ = 2.5‰), indicating that they were formed from evaporatively concentrated brines after deposition. These data together suggest that the HFS L2.1 porosity trend may have been formed by early marine calcite cementation in well-agitated and well-oxygenated conditions along the ramp crest. The lower porosity in updip dolomites is probably the result of predolomitization compaction and possibly porosity occlusion by dolomite and anhydrite cementation. Both compaction and dolomitization, and thus, porosity loss, were probably limited in the ramp crest because of early calcite cementation and stabilization that occurred there.

History of Reservoir Development

The evolution of the Fullerton reservoir succession can be summarized as follows. During the late L1 to early L2, most of the Fullerton area was an area of low-accommodation, peritidal, tidal-flat deposition (Wichita facies). During periodic exposure of this shallow-water platform (by relative fall of sea level), these facies underwent early dolomitization and stabilization of mineralogy and porosity by dominantly

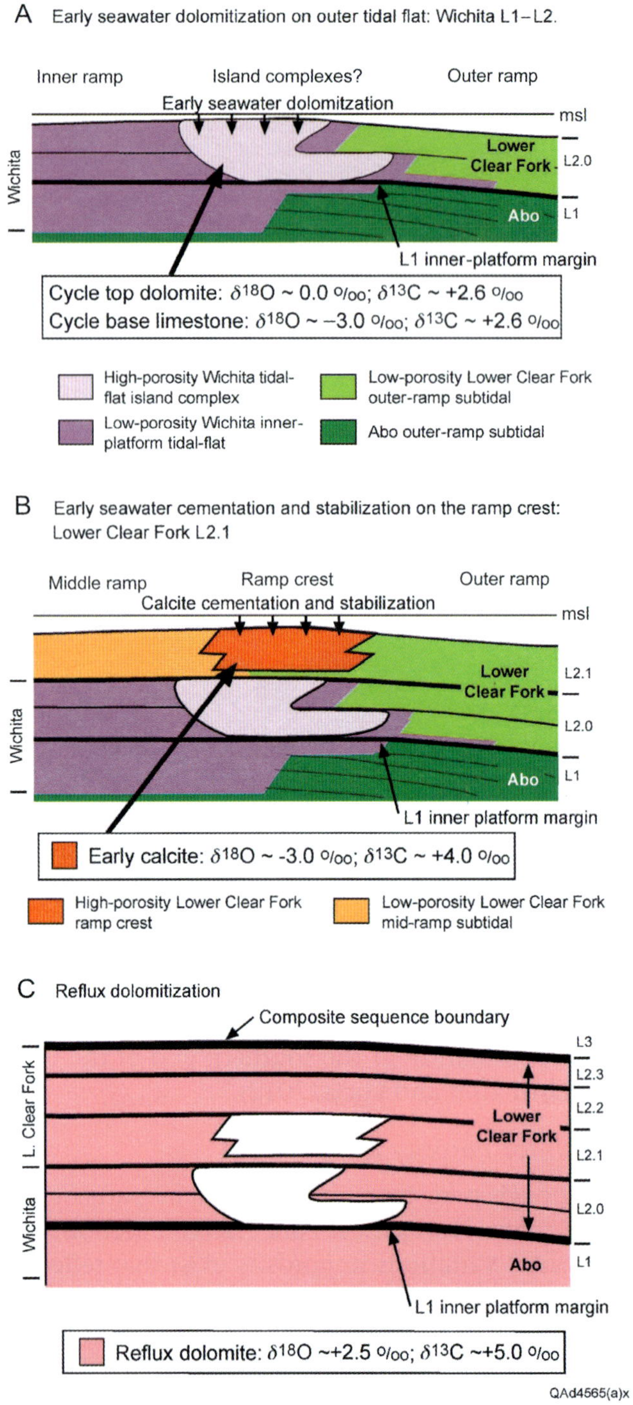

FIGURE 31. Models for diagenesis and porosity formation at Fullerton field. (A) Early seawater dolomitization on a tidal flat; Wichita, L1–L2. Wichita porosity development associated with early seawater dolomitization along the outer margins of tidal-flat complex. (B) Early seawater cementation and stabilization on a ramp crest; Lower Clear Fork, L2.1. Porosity related to the preservation of primary porosity in Lower Clear Fork ramp-crest limestones. (C) Reflux dolomitization at sequence boundaries. Most of the Wichita and the Lower Clear Fork was dolomitized by evaporatively concentrated reflux brines. See text for discussion. msl = mean sea level. Modified from Ruppel and Jones (2006).

marine waters. Isotopic data suggest that this diagenesis focused along the outer margins of the platform in what may have been a tidal-flat island complex (Figure 31A). More landward areas of the tidal flat were not dolomitized at this time and lost porosity through compaction and porosity occlusion.

During subsequent transgression of the platform in L2.1, subtidal middle-platform (Lower Clear Fork) facies tracts were established across the area (Figure 15). Early mineralogical stabilization of high-energy grain-rich sediments occurred along the ramp crest, which also paralleled the platform margin (Figure 31B). This early stabilization served to preserve both the original seawater chemistry of these rocks and much of the primary porosity.

Isotope data suggest that the remainder of the unstabilized reservoir succession (Wichita–Lower Clear Fork) was later dolomitized by evaporatively concentrated reflux brines (Figure 31C). The exact timing of this diagenesis, which resulted in dolomitization of all parts of the succession except for early dolomitized L1 (Wichita) tidal-flat areas and the early stabilized L2 (Lower Clear Fork) ramp crest, is poorly constrained. Stratigraphic relationships and strontium isotope data (Ruppel and Jones, 2006) suggest that dolomitizing reflux brines were most likely generated during L2, presumably at one or more sequence boundaries, including the top of the L2-L3 sequence boundary.

SUMMARY AND CONCLUSIONS

The procedures used to develop the reservoir framework at Fullerton field and many attributes of this framework offer important guidelines for the characterization of Clear Fork reservoirs and for the development of predictive models of the distribution of reservoir properties. Key findings from this study include the following.

The Leonardian reservoir succession in the Permian Basin consists of three formations or facies successions (from oldest to youngest): Abo, Wichita, and Lower Clear Fork. Each of these formations is characterized by distinctive facies, cyclicity, depositional architecture, and porosity development.

The Abo consists mostly of porous outer-ramp, open-marine, fusulinid-crinoid facies whose clinoformal architecture is clearly expressed on seismic data. Lateral facies continuity in these rocks, which represent all but the most proximal facies of the basal Leonardian (L1) sequence, is poor. As a result, porosity distribution is complex and probably highly discontinuous.

The Wichita, which includes both a lower section that is part of the L1 HST and an upper section that is part of the L2 TST, consists of a thick succession of aggradational, restricted, tidal-flat facies. These rocks contain high porosity but commonly relatively low permeability and display poor small-scale continuity. Cyclicity is poorly developed; reservoir architecture is more controlled by diagenesis than by depositional facies. Limestone intervals in the dominantly dolomitized Wichita display extremely low porosity and permeability and act as local fluid-flow baffles between the higher porosity intervals of dolostone. Karst features (including inclined beds and monomict [Figure 20B–E] and polymict cave-fill breccias) are common in the Wichita. Neither wireline-log data nor core-analysis data reveal any definitive differences between karsted and unkarsted sections of Wichita. Nevertheless, these karst zones, which are definable only on image logs and cores, may locally affect fluid flow.

The Lower Clear Fork consists of three HFSs within sequence L2, each recording sea level rise (transgression) and fall (regression). Reservoir development is restricted mostly to subtidal facies (late transgression and early highstand). The highest porosity and permeability in the Lower Clear Fork are associated with incompletely dolomitized grain-rich packstones and grainstones. Cyclicity is well developed. However, facies are discontinuous at the cycle scale.

Gamma-ray logs are useful for general correlations only. As in other Leonardian reservoirs, a high gamma-ray response is commonly an indication of the presence of clays and silt in tidal-flat facies; a low gamma-ray response indicates subtidal facies. Porosity logs or image logs must be used for high-resolution cycle-scale correlation. Three-dimensional seismic provides excellent resolution of both reservoir architecture and distribution of reservoir porosity at the HFS scale.

Porosity development is mostly a function of early diagenesis controlled by platform-margin geometry. Zones of highest porosity in the Wichita are the result of early dolomitization and stabilization along the outer margin of the Wichita inner platform. A similar subparallel zone in the Lower Clear Fork was created by early calcite cementation and mineralogical stabilization in the platform-ramp crest.

ACKNOWLEDGMENTS

The results presented in this chapter are part of a continuing research into the styles and causes of heterogeneity in shallow-water platform-carbonate reservoirs in the Permian Basin by the Bureau of Economic Geology (BEG). Discussions with colleagues at BEG, including Charlie Kerans, Robert Loucks, Jerry Lucia, and Jim Jennings, were especially helpful in formulating the interpretations presented herein. The funding for the study was provided by contract no. DE-FC26-01BC15351 from the U.S. Department of Energy. Additional funding for the study was provided by ExxonMobil and The University of Texas System. We thank David Smith, Terry Anthony, Steve Krohn, and Amy Powell of ExxonMobil, and Jeff Simmons, Craig Kemp, and John Stout of Oxy, for their contributions in organizing the project and in providing data. David Smith and Terry Anthony were especially helpful in providing insights into field geology and engineering issues. We also thank Stephen Hartman and Tim Hunt of The University of Texas System, West Texas Operations Office, for providing both collaborative funding and data. Ursula Hammes and Wayne Wright provided useful critical review comments that improved the manuscript.

REFERENCES CITED

Atchley, S. C., M. G. Kozar, and L. A. Yose, 1999, A predictive model for reservoir characterization in the Permian (Leonardian) Clear Fork and Glorieta formations, Robertson field area, west Texas: AAPG Bulletin, v. 83, p. 1031–1056.

Bebout, D. G., F. J. Lucia, C. R. Hocott, G. E. Fogg, and G. W. Vander Stoep, 1987, Characterization of the Grayburg reservoir, University Lands Dune field, Crane County, Texas: Austin, Texas, University of Texas at Austin, Bureau of Economic Geology Report of Investigations 168, 98 p.

Dutton, S. P., E. M. Kim, R. F. Broadhead, C. L. Breton, W. D. Raatz, S. C. Ruppel, and C. Kerans, 2004, Play analysis and digital portfolio of major oil reservoirs in the Permian Basin: Application and transfer of advanced geological and engineering technologies for incremental production opportunities: Austin, Texas, University of Texas at Austin, Bureau of Economic Geology, final report prepared for the Department of Energy under contract no. DE-FC26-02NT15131, 408 p.

Fitchen, W. M., M. A. Starcher, R. T. Buffler, and G. L. Wilde, 1995, Sequence-stratigraphic framework and facies models of the early Permian platform margins, Sierra Diablo, west Texas, *in* R. A. Garber and R. F. Lindsay, eds., Wolfcampian–Leonardian shelf-margin facies of the Sierra Diablo: Seismic scale models for subsurface exploration: West Texas Geological Society Publication 95–97, p. 23–66.

Garber, R. A., and P. M. Harris, 1990, Depositional facies of the Grayburg/San Andres dolomite reservoirs, Central Basin platform, Permian Basin, *in* D. G. Bebout and P. M. Harris, eds., Geologic and engineering approaches in evaluation of San Andres/Grayburg hydrocarbon reservoirs: Permian Basin: Austin, Texas, University of Texas at Austin, Bureau of Economic Geology, p. 1–20.

Hammes, U., 1997, Electrical imaging catalog: Microresistivity images and core photos from fractured, karsted, and brecciated carbonate rocks: Austin, Texas, University of Texas at Austin, Bureau of Economic Geology Geological Circular 97-2, 40 p.

Harrington, R. R., and F. J. Lucia, 2012, Integration of rock fabrics and stratigraphy for petrophysical quantification of reservoir framework, *in* S. C. Ruppel, ed., Anatomy of a giant carbonate reservoir: Fullerton Clear Fork (Lower Permian) field, Permian Basin, Texas: Studies in Geology 63, p. 49–66.

Holtz, M. H., S. C. Ruppel, and C. R. Hocott, 1992, Integrated geologic and engineering determination of oil-reserve growth potential in carbonate reservoirs: Journal of Petroleum Technology, v. 44, p. 1250–1258.

Jones, R., and S. Ruppel, 2004, Evidence of post-Wolfcampian fault movement and its impact on Clear Fork reservoir quality: Fullerton field, west Texas, *in* R. C. Trentham, ed., Banking on the Permian Basin: Plays, field studies, and techniques: West Texas Geological Society Fall Symposium: West Texas Geological Society Publication 04-112, p. 207.

Kerans, C., and W. M. Fitchen, 1995, Sequence hierarchy and facies architecture of a carbonate ramp system: San Andres Formation of Algerita escarpment and western Guadalupe Mountains, west Texas and New Mexico: Austin, Texas, University of Texas at Austin, Bureau of Economic Geology Report of Investigations 235, 86 p.

Kerans, C., and K. Kempter, 2002, Hierarchical stratigraphic analysis of a carbonate platform, Permian of the Guadalupe Mountains: Austin, Texas, University of Texas at Austin, Bureau of Economic Geology (AAPG/Datapages Discovery Series 5), CD-ROM.

Kerans, C., and S. C. Ruppel, 1994, San Andres sequence framework, Guadalupe Mountains: Implications for San Andres type section and subsurface reservoirs, *in* R. A. Garber and D. R. Keller, eds., Field guide to the Paleozoic section of the San Andres Mountains: Permian Basin Section: SEPM Publication 94-35, p. 105–115.

Kerans, C., F. J. Lucia, and R. K. Senger, 1994, Integrated characterization of carbonate ramp reservoirs using Permian San Andres Formation outcrop analogs: AAPG Bulletin, v. 78, no. 2, p. 181–216.

Kerans, C., K. Kempter, J. Rush, and W. L. Fisher, 2000, Facies and stratigraphic controls on a coastal paleokarst: Lower Permian, Apache Canyon, west Texas, *in* R. Lindsay, R. Trentham, R. F. Ward, and A. H. Smith, eds., Classic Permian geology of west Texas and southeastern New Mexico: 75 years of Permian Basin oil and gas exploration and development: West Texas Geological Society Publication 00-108, p. 55–82.

King, P. B., 1942, Permian of west Texas and southeastern New Mexico: AAPG Bulletin, v. 26, no. 4, p. 535–763.

King, P. B., 1965, Geology of the Sierra Diablo region, Texas: U.S. Geological Survey Professional Paper 480, 185 p.

Kluth, C. F., and P. J. Coney, 1981, Plate tectonics of the Ancestral Rockies: Geology, v. 9, p. 10–15, doi:10.1130/0091-7613(1981)9<10:PTOTAR>2.0.CO;2.

Lohman, K. C., and J. C. G. Walker, 1989, The δ^{18}O record of Phanerozoic abiotic cements: Geophysical Research Letters, v. 16, p. 319–322, doi:10.1029/GL016i004p00319.

Longacre, S. A., 1990, The Grayburg reservoir, North McElroy Unit, Crane County, Texas, *in* D. G. Bebout and P. M. Harris, eds., Geologic and engineering approaches in evaluation of San Andres/Grayburg hydrocarbon reservoirs: Permian Basin: Austin, Texas, University of Texas at Austin, Bureau of Economic Geology, p. 239–273.

Loucks, R. G., 1999, Paleocave carbonate reservoirs: Origins, burial-depth modifications, spatial complexity, and reservoir implications: AAPG Bulletin, v. 83, p. 1795–1834.

Loucks, R. G., and P. Mescher, 2001, Paleocave facies classification and associated pore types: AAPG Southwest Section Annual Meeting, Dallas, Texas, March 11–13, CD-ROM, 18 p.

Lucia, F. J., and J. W. Jennings Jr., 2002, Calculation and distribution of petrophysical properties in the south Wasson Clear Fork field, *in* F. J. Lucia, ed., Integrated outcrop and subsurface studies of the interwell environment of carbonate reservoirs: Clear Fork (Leonardian-age) reservoirs, west Texas and New Mexico, final technical report prepared for the U.S. Department of Energy under contract no. DE-AC26-98BC15105, p. 95–142.

Lucia, F. J., and J. A. Kane, 2012, Calculation of permeability and initial water saturations from wireline logs in a mature carbonate reservoir, *in* S. C. Ruppel, ed., Anatomy of a giant carbonate reservoir: Fullerton Clear Fork (Lower Permian) field, Permian Basin, Texas: Studies in Geology 63, p. 77–91.

Major, R. P., G. W. Vander Stoep, and M. H. Holtz, 1990, Delineation of unrecovered mobile oil in a mature dolomite reservoir: East Penwell San Andres Unit, University Lands, west Texas: Austin, Texas, University of Texas at Austin, Bureau of Economic Geology Report of Investigations 194, 52 p.

Mazzullo, S. J., 1982, Stratigraphy and depositional mosaics of Lower Clear Fork and Wichita groups (Permian), northern Midland Basin, Texas: AAPG Bulletin, v. 66, p. 210–227.

Mazzullo, S. J., 1994, Dolomitization of periplatform carbonates (Lower Permian, Leonardian), Midland Basin, Texas: Carbonates and evaporites, v. 9, no. 1, p. 95–112, doi:10.1007/BF03175189.

Mazzullo, S. J., and A. Reid, 1989, Lower Permian platform and basin depositional systems, northern Midland Basin, Texas, *in* P. D. Crevello, J. J. Wilson, J. F. Sarg, and J. F. Read, eds., Controls on carbonate platform and basin development: SEPM Special Publication, v. 44, p. 305–320.

Presley, M. W., 1987, Evolution of Permian evaporite basin in Texas panhandle: AAPG Bulletin, v. 71, p. 167–190.

Presley, M. W., and K. A. McGillis, 1982, Coastal evaporite and tidal-flat sediments of the Upper Clear Fork and Glorieta formations, Texas panhandle: Austin, Texas, University of Texas at Austin, Bureau of Economic Geology Report of Investigations 115, 50 p.

Ruppel, S. C., 1992, Styles of deposition and diagenesis in Leonardian carbonate reservoirs in west Texas: Implications for improved reservoir characterization: Society of Petroleum Engineers Annual Exhibition and Technical Conference, Washington, D. C., October 4–7, 1992, SPE Paper 24691, p. 313–320.

Ruppel, S. C., 2002, Geological controls on reservoir development in a Leonardian (Lower Permian) carbonate platform reservoir, Monahans field, west Texas: Austin, Texas, University of Texas at Austin, Bureau of Economic Geology Report of Investigations 266, 58 p.

Ruppel, S. C., and E. E. Ariza, 2002, Cycle and sequence stratigraphy of the Clear Fork reservoir at South Wasson field, Gaines County, Texas, *in* F. J. Lucia, ed., Integrated outcrop and subsurface studies of the interwell environment of carbonate reservoirs: Clear Fork (Leonardian-age) reservoirs, west Texas and New Mexico, final technical report prepared for the U.S. Department of Energy under contract no. DE-AC26-98BC15105, p. 59–94.

Ruppel, S. C., and D. G. Bebout, 2001, Competing effects of depositional architecture and diagenesis on carbonate reservoir development: Grayburg Formation, South Cowden field, west Texas: Austin, Texas, University of Texas at Austin, Bureau of Economic Geology Report of Investigations 263, 62 p.

Ruppel, S. C., and H. S. Cander, 1988a, Effects of facies and diagenesis on reservoir heterogeneity: Emma San Andres field, west Texas: Austin, Texas, University of Texas at Austin, Bureau of Economic Geology Report of Investigations 178, 67 p.

Ruppel, S. C., and H. S. Cander, 1988b, Dolomitization of shallow-water platform carbonates by seawater and seawater-derived brines: San Andres Formation (Guadalupian), west Texas, *in* Sedimentology and geochemistry of dolostones: SEPM Special Publication 43, p. 245–262.

Ruppel, S. C., and R. H. Jones, 2006, Key role of outcrops and cores in carbonate reservoir characterization and modeling, Lower Permian Fullerton field, Permian Basin, United States, *in* P. M. Harris and L. J. Weber, eds., Giant hydrocarbon reservoirs of the world: From rocks to reservoir characterization and modeling: AAPG Memoir 88, p. 355–394.

Ruppel, S. C., W. B. Ward, E. E. Ariza, and J. W. Jennings Jr., 2000, Cycle and sequence stratigraphy of Clear Fork reservoir-equivalent outcrops: Victorio Peak Formation, Sierra Diablo, Texas, *in* R. Lindsay, R. Trentham, R. F. Ward, and A. H. Smith, eds., Classic Permian geology of west Texas and southeastern New Mexico: 75 years of Permian Basin oil and gas exploration and development: West Texas Geological Society Publication 00-108, p. 109–130.

Saller, A. H., and N. Henderson, 1998, Distribution of porosity and permeability in platform dolomites: Insights from the Permian of west Texas: AAPG Bulletin, v. 82, no. 8, p. 1528–1550.

Tyler, N., and N. J. Banta, 1989, Oil and gas resources remaining in the Permian Basin: Targets for additional hydrocarbon recovery: Austin, Texas, University of Texas at Austin, Bureau of Economic Geology Geological Circular 89-4, 20 p.

Vogt, J. N., 1986, Dolomitization and anhydrite diagenesis of the San Andres (Permian) Formation, Gaines County, Texas: Master's thesis, University of Texas at Austin, Austin, Texas, 202 p.

Wang, F. P., and F. J. Lucia, 2012, Reservoir modeling and simulation of the Fullerton Clear Fork reservoir, Andrews County, Texas, *in* S. C. Ruppel, ed., Anatomy of a giant carbonate reservoir: Fullerton Clear Fork (Lower Permian) field, Permian Basin, Texas: Studies in Geology 63, p. 111–143.

Ye, Q., and S. J. Mazzullo, 1993, Dolomitization of lower Permian platform facies, Wichita Formation, north platform, Midland Basin, Texas: Carbonates and evaporites, v. 8, no. 1, p. 55–70, doi:10.1007/BF03175163.

Zeng, H., 2012, Construction and analysis of three-dimensional seismic porosity inversion models, *in* S. C. Ruppel, ed., Anatomy of a giant carbonate reservoir: Fullerton Clear Fork (Lower Permian) field, Permian Basin, Texas: Studies in Geology 63, p. 93–110.

Zeng, H., and C. Kerans, 2003, Seismic frequency control on carbonate seismic stratigraphy: A case study of the Kingdom Abo sequence, west Texas: AAPG Bulletin, v. 87, no. 2, p. 273–293, doi:10.1306/08270201023.

2

Harrington, R. R., and F. J. Lucia, 2012, Integration of rock fabrics and stratigraphy for petrophysical quantification of reservoir framework, *in* S. C. Ruppel, ed., Anatomy of a giant carbonate reservoir: Fullerton Clear Fork (Lower Permian) field, Permian Basin, Texas: Studies in Geology 63, p. 49–66.

Integration of Rock Fabrics and Stratigraphy for Petrophysical Quantification of Reservoir Framework

Rebecca R. Harrington[1]
Bureau of Economic Geology, The University of Texas at Austin, Austin, Texas, U.S.A.

F. Jerry Lucia
Bureau of Economic Geology, The University of Texas at Austin, Austin, Texas, U.S.A.

ABSTRACT

A major task in building a reservoir model is quantifying the geologic framework using petrophysical properties. Porosity and water saturation values can be obtained from wireline logs, but permeability is a rock property that cannot be obtained directly from logs. Commonly, a single porosity-permeability transform is used to estimate permeability from porosity logs. However, such an approach fails to account for variations in porosity-permeability relationships that are common in carbonates. In this study, rock-fabric-specific porosity-permeability transforms are used together with log porosity to calculate permeability. Although the reservoir at Fullerton field contains several rock fabrics, they can all be grouped into three petrophysical groups, each having a unique porosity-permeability relationship. These petrophysical groups can be linked to facies and stratigraphy using an integrated study of thin sections and cores.

The lowermost stratigraphic unit in the reservoir (the Wichita of sequences L1 and L2) is composed dominantly of peritidal facies consisting of fine-crystalline mud-dominated dolostones and mud-dominated limestones, all of which can be assigned to a single petrophysical group (class 3). In contrast, rocks of the overlying Lower Clear Fork (sequences L2.1–L2.3) display far more petrophysical variability, both stratigraphically and geographically. Sequence L2.1 comprises an upper late

[1] *Present address*: XTO Energy, Fort Worth, Texas, U.S.A.

DOI:10.1306/13341537St633525

highstand peritidal succession and a lower, transgressive, and early highstand systems tract succession of subtidal facies. The peritidal rocks such as those of the Wichita are fine-crystalline mud-dominated fabrics of petrophysical class 3. The subtidal rocks, which contain both limestone and dolostone, are more variable. Dolostones are mostly medium-crystalline, subtidal grain-dominated dolopackstones and medium-crystalline, mud-dominated petrophysical class 2 rocks. Limestones are composed of oomoldic grainstone. Although grainstones normally possess petrophysical class 1 rock fabrics, they display class 2 petrophysical relationships because they are moldic. Lower Clear Fork sequence L2.2 is also dominated by class 2 medium-crystalline dolostone fabrics. However, these rocks display class 1 porosity-permeability relationships because of abundant poikilotopic anhydrite. Oomoldic limestone, such as that in sequence L2.1, is also locally common. The uppermost L2.3 reservoir sequence contains fine-crystalline dolostone, class 3 peritidal facies.

INTRODUCTION

The theory and development of the rock-fabric approach for quantifying geologic models were first described by the Bureau of Economic Geology (BEG) researchers Kerans, Lucia, and Senger (Kerans et al., 1994; Lucia, 1995, 2007; Jennings and Lucia, 2001). This approach has been successfully applied to several other Permian Basin carbonate reservoirs, including the San Andres at Seminole field (Lucia et al., 1995; Wang et al., 1998), the Clear Fork at South Wasson field (Lucia and Jennings, 2002), and the Grayburg at South Cowden field (Lucia and Ruppel, 1996; Ruppel and Lucia, 1996; Lucia, 2000). The goal of the rock-fabric approach is to better define permeability and original water saturation distribution in carbonate reservoirs. Permeability, unlike porosity and water saturation, cannot be measured by wireline devices and therefore is not easily obtained.

Research at BEG has shown that permeability and water saturation are functions of rock fabric and interparticle porosity. The three basic rock fabrics in limestones are grainstone or dolograinstone, grain-dominated packstone or dolopackstone, and mud-dominated rocks composed of mud-dominated packstone, wackestone, and mudstone. Each of these rock-fabric classes is characterized by different relationships between porosity and permeability that allow them to be segregated into petrophysical classes (Lucia, 1995) (Figure 1A). Each petrophysical class is characterized by a specific porosity-permeability transform (Figure 1B), assuming that only interparticle pore space is present. Petrophysical class 1 rock fabrics include grainstones, dolograinstones, and coarse-crystalline (>100 μm) dolostones (Figure 2); class 2 fabrics include grain-dominated packstones, grain-dominated dolopackstones, and medium-crystalline (20–100 μm) dolostones (Figure 3); class 3 fabrics include mud-dominated packstones, wackestones, and mudstones and their fine-crystalline (<20 μm), dolomitized counterparts (Figure 4). Note that rocks of different petrophysical classes exhibit very different petrophysical relationships. For example, class 3 rocks having 10% porosity typically display about 2-md permeability, whereas class 1 rocks with the same porosity would typically exhibit about 20-md permeability (Figure 1B).

Previous studies (Lucia, 1995; Wang et al., 1998; Lucia and Jennings, 2002) have shown that, in some cases, permeability can be estimated from wireline logs if data are sufficient to determine interparticle porosity and rock fabrics. However, a complete suite of wireline logs is commonly necessary for these calculations. In Fullerton Clear Fork field, few suites of wireline logs are complete, and the reservoir has been under waterflood for more than 40 yr, further complicating log-based calculations. Consequently, relative proportions of separate-vug and interparticle porosity and, therefore, petrophysical class, could not be determined from logs.

Estimating the permeability from porosity logs in this reservoir required a more innovative approach. Distributions of rock fabrics were determined from thin-section descriptions of samples from each stratigraphic interval. Rock-fabric data were then used to assign appropriate petrophysical transforms throughout the field. This approach is based on the concept that petrophysical class is a product of rock fabrics and that this link between geologic description and petrophysical properties should result in a predictable relationship between petrophysical properties and stratigraphy. This chapter describes the development of this relationship.

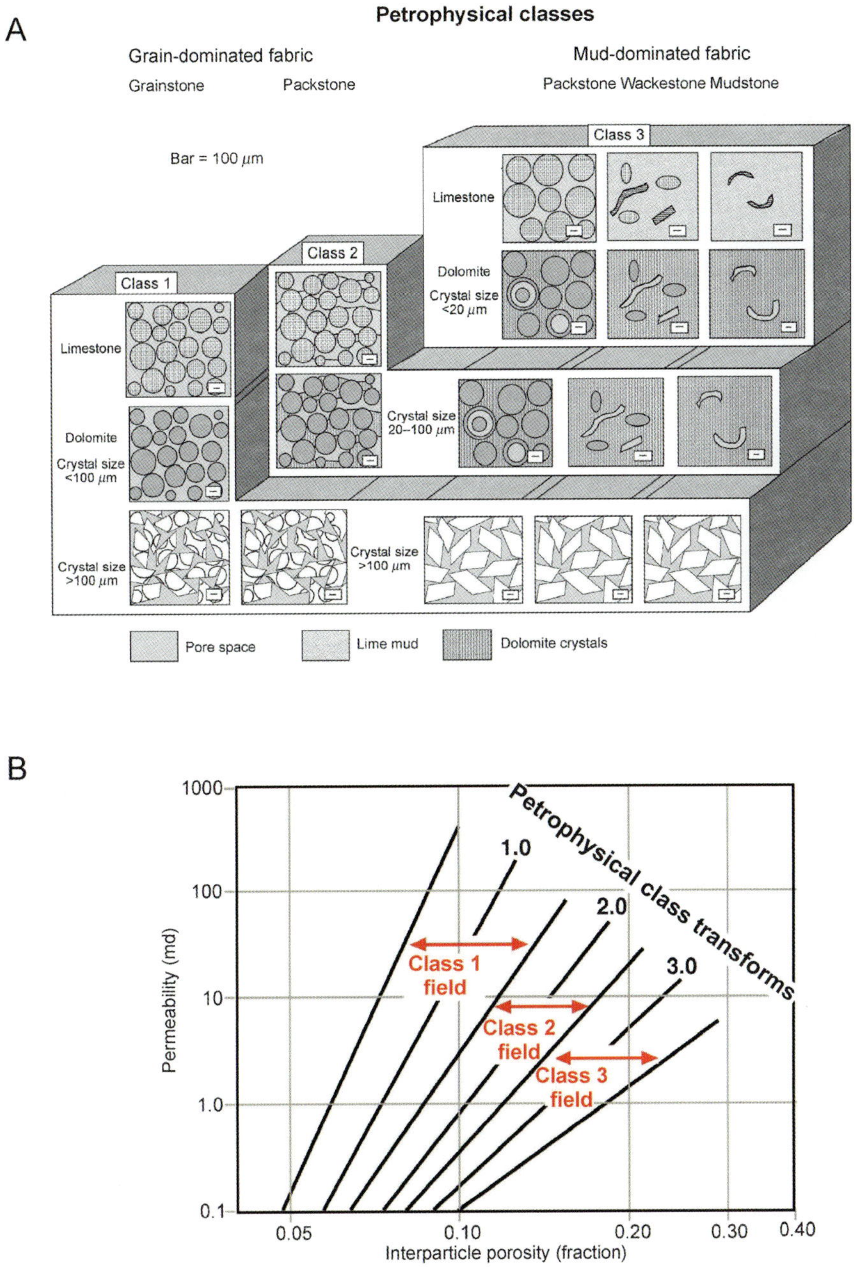

FIGURE 1. (A) Diagram showing rock fabrics composing petrophysical classes 1, 2, and 3. (B) A crossplot showing interparticle porosity versus permeability, petrophysical class fields, and associated transforms.

METHODS

Initial rock-fabric descriptions were done in an area of abundant high-quality core and log data (Figure 5). Two cores from this area were resampled for porosity-permeability analysis and thin-section preparation throughout the productive interval, including 384 plugs and sections from Fullerton Clear Fork Unit (FCU) 5927 and 111 from FCU 6429. New sample plugs were cut from every foot of core and analyzed for porosity and permeability. Thin sections were made from a wafer sliced from the end of each plug. Matching thin sections and core analyses were also obtained similarly from 210 samples from FCU 7630 in another part of the field (Figure 5). New thin sections (245) were prepared from several other cores in the north-central part of the field (FCU 6429, FCU 6229, FCU 6122) and in the southern part of the field (FCU 4828, University Consolidated [Univ. Cons.] IV-25) to refine rock-fabric interpretations.

Existing petrophysical data sets and thin sections were used to fill in areas between control wells. Data from 787 existing thin sections from 14 wells and whole-core analyses from 63 wells were used. Neutron-density and photoelectric factor logs from 108 wells were used to determine mineralogy for mapping lithology.

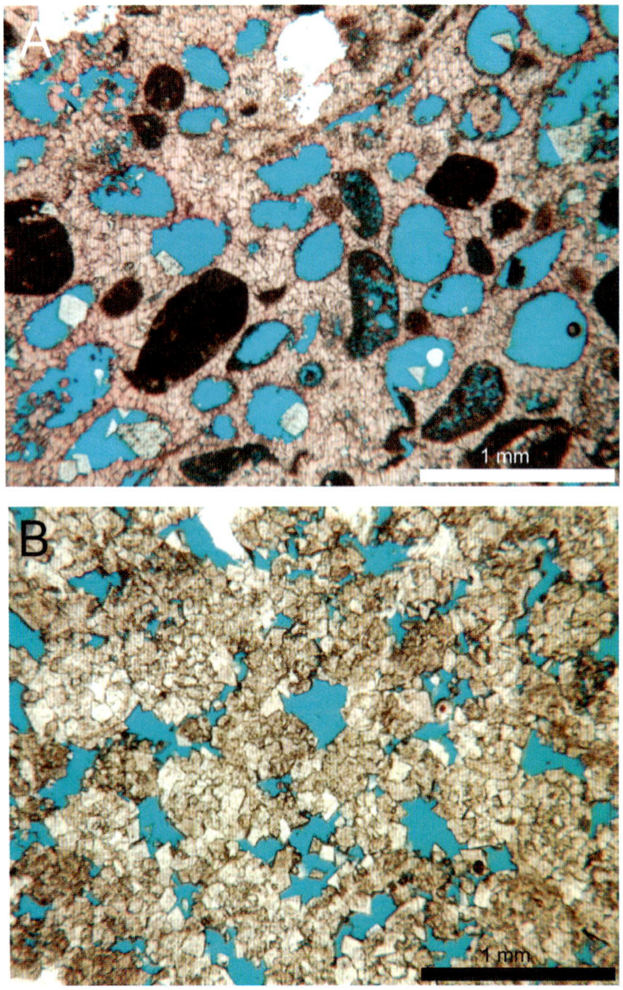

FIGURE 2. **Photomicrographs of class 1 fabrics. (A) Lime grainstone with separate-vug pore space: FCU 5927′, 6913′, ϕ = 22.0%, k = 2.69 md. (B) Coarse-crystalline dolograinstone: FCU 7630′, 6997′, ϕ = 11.7%, k = 112.56 md. Dolomite crystal size = 120 μm. FCU = Fullerton Clear Fork Unit.**

Thin sections were described at BEG on a polarizing transmitted-light microscope. Visual mineralogy estimates were made of percentages of calcite, dolomite, and anhydrite. Trace amounts of pyrite and quartz were also observed. As a percentage of the carbonate fraction, dolostones were defined as containing 80% or more dolomite, and limestones, as containing less than 80% dolomite. Dolomite crystal size (in micrometers) was measured by selecting an average-size crystal and measuring the distance between its opposing crystal faces. Fabrics were described using the Lucia classification system (Lucia, 1995). The fabrics include grainstone, grain-dominated packstone, mud-dominated packstone, wackestone, and mudstone. Because tidal-flat facies typically contain separate-vug pore types, for example, fenestral, they were not easily compared with interparticle pore systems and were thus classified empirically. Interparticle, separate-vug, and touching-vug porosity, including fractures, were also estimated. Point counts (300 per section) were performed as necessary using a mechanical stage in areas of complex lithology or pore-type distribution to improve accuracy. Appendix I in this chapter contains a detailed description of procedures used for describing thin sections using the rock-fabric approach.

RESULTS

Rock-fabric variations were observed on many scales in the north-central part of the field. Lower Clear Fork cycles typically are composed of mud-dominated facies at their bases and grain-dominated facies at their tops (Ruppel and Harrington, 2012). Normally, these

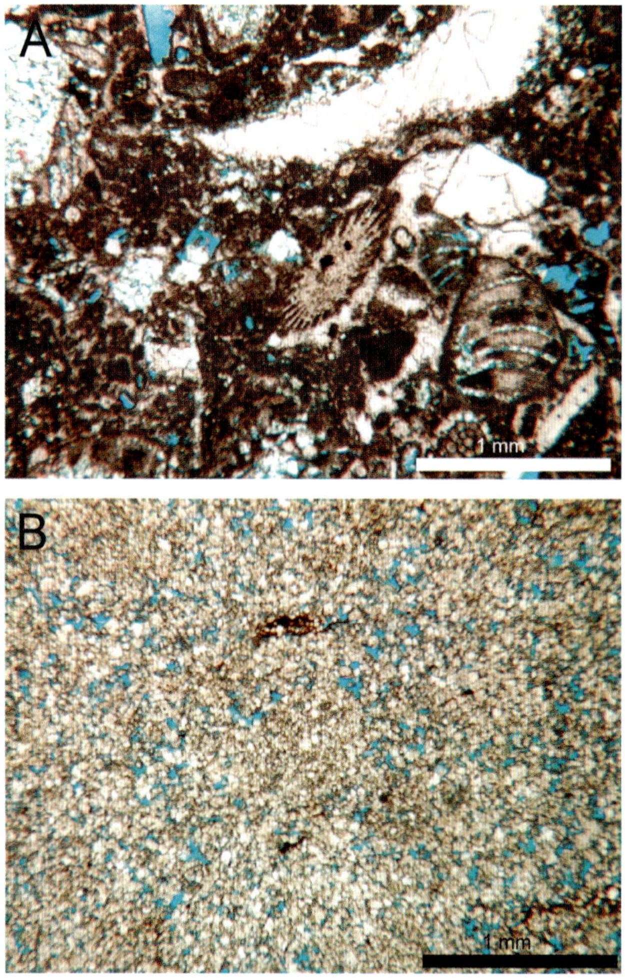

FIGURE 3. **Photomicrographs of class 2 fabrics. (A) Grain-dominated packstone with separate-vug pore space: FCU 5927′, 6987′, ϕ = 8.8%, k = 0.04 md. (B) Medium-crystalline, grain-dominated dolopackstone: FCU 6429′, 6914′, ϕ = 18.1%, k = 27.2 md. Dolomite crystal size = 55 μm. FCU = Fullerton Clear Fork Unit.**

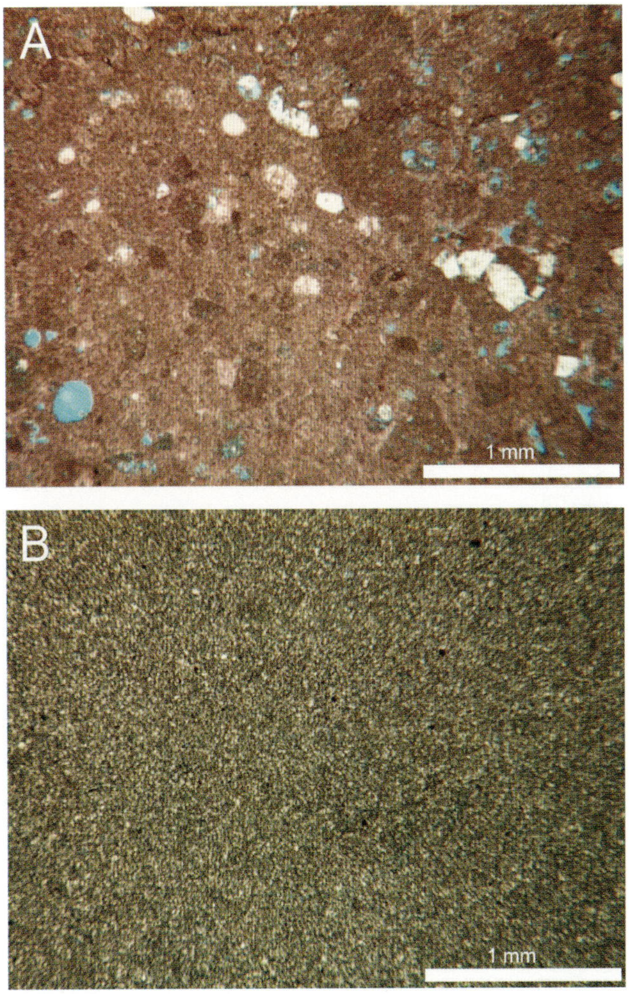

FIGURE 4. Photomicrographs of class 3 fabrics. (A) Wackestone: FCU 5927′, 6903′, $\phi = 11.2\%$, $k = 0.74$ md. (B) Fine-crystalline dolomudstone with microporosity: FCU 5927′, 7009′, $\phi = 15.7\%$, $k = 0.86$ md. Dolomite crystal size = less than 10 μm. FCU = Fullerton Clear Fork Unit.

facies relationships would imply class 3 petrophysical properties at cycle bases and class 2 or 1 properties at cycle tops. Our detailed rock-fabric and petrophysical studies at Fullerton showed, however, that both cycle-base mud-rich rocks and grain-rich cycle-top rocks exhibit class 2 relationships. This occurs because mud-rich facies have been altered during diagenesis to medium-crystal dolomudstones that have class 2 petrophysics like the grain-dominated dolopackstones at cycle tops. Wichita cycles typically are composed of mud-dominated fabrics capped by tidal-flat facies. Both facies are fine-crystalline dolomites and representative of a petrophysical class. Peritidal rocks in the Lower Clear Fork are also assigned to class 3.

These observations initially suggested that a simple stratigraphic model could be used to determine petrophysical class (and permeability) across the entire reservoir using porosity wireline logs: a class 3 transform for Wichita rocks and peritidal Lower Clear Fork rocks and a class 2 transform for Lower Clear Fork subtidal rocks. Further study, however, revealed both lateral and vertical variations, indicating that fieldwide relationships are more complex. The stratigraphic and geographic variations in rock fabrics and petrophysical class are discussed by formation in the following sections in this chapter. A vertical one-dimensional profile depicting basic stratigraphic variations in petrophysical class for the reservoir is shown in Figure 6.

Abo Formation

The Abo Formation was not studied extensively because it is not part of the main reservoir (Ruppel and Harrington, 2012). However, cores and core-analysis data in the field are sufficient to suggest that this formation can be characterized as a class 1 coarse-crystalline dolostone. No new core analyses or thin sections were obtained for the Abo, but existing data plot mostly as class 1 (Figure 7). Although a more detailed study of the Abo would most likely reveal more variation, a petrophysical class 1 transform was assigned to the Abo Formation throughout the field for estimating permeability and initial water saturation.

Wichita Formation

Ruppel and Harrington (2012) showed that the Wichita Formation represents proximal facies tracts of two depositional sequences. The lower Wichita is an inner-platform equivalent of the Abo outer-platform succession. Both were deposited during late highstand of sequence L1. The upper Wichita is the proximal equivalent of the lower part of the Lower Clear Fork. These rocks accumulated as basal transgressive deposits of sequence L2.

Wichita rocks consist of mud-dominated fabrics, including mudstones, wackestones, mud-dominated packstones, and tidal-flat fabrics. Typical tidal-flat features include pisolites, fenestrae, fine laminations, and desiccation cracks (Figure 8). Fenestrae are commonly lined with dolomite crystals and filled with anhydrite or, less commonly, calcite. Fractures are locally present and typically filled with anhydrite or calcite. Very low porosity (<3%) limestones are interbedded with porous dolostones in the north-central part of the field (Figure 9) and may act locally as baffles to fluid flow. The rest of the Wichita contains primarily fine-crystalline dolostones. Both limestones

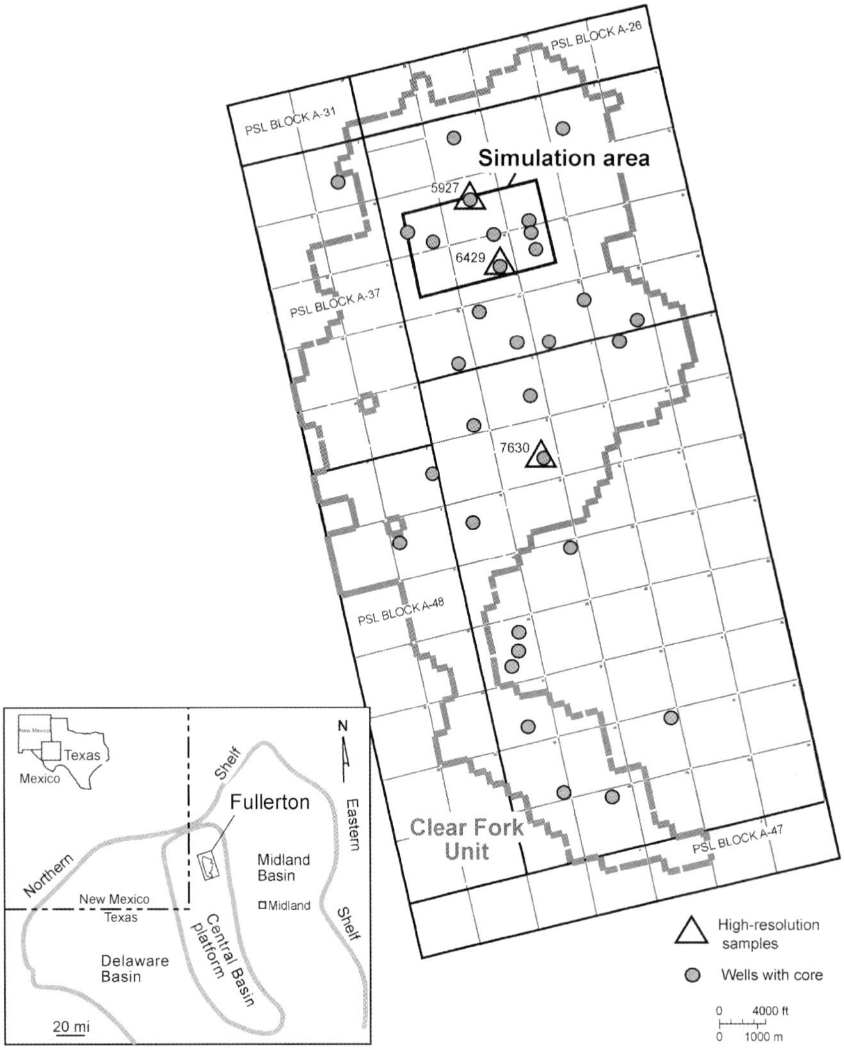

FIGURE 5. Map showing the location of Fullerton field, Andrews County, Texas, and locations of flow simulation study area and cored wells. PSL = public school lands.

and dolostones are dominantly petrophysical class 3 because of their high mud content and fine crystal size (Figure 10). Class 2 grain-dominated dolopackstones and medium-crystalline mud-dominated dolostones are locally present but are uncommon in most of the field. Therefore, the formation was assigned petrophysical class 3, and class 3 transforms were used to estimate permeability and initial water saturation.

Lower Clear Fork Formation

Most of the Lower Clear Fork (sequences L2.0, L2.1, and L2.2) is composed primarily of subtidal facies with minor tidal-flat facies. Subtidal facies contain skeletal debris, including fusulinids, peloids, and ooids (Ruppel and Harrington, 2012). Rock fabrics, which include both limestones and dolostones, comprise wackestone, mud-dominated packstone, grain-dominated packstone, and grainstone. Limestones are grain-dominated (Figures 2A, 3A) and commonly more porous than dolostones, with most porosity values ranging from 5 to 15% (locally as high as 30%). These subtidal rocks are dominantly of petrophysical classes 1 and 2 because of the more grain-rich textures of the limestones and the medium and coarse dolomite crystal size in the dolostones.

Lower Clear Fork L2.0

In most of the field area, high-frequency sequence (HFS) L2.0 consists of amalgamated tidal-flat deposits of the upper Wichita (Ruppel and Harrington, 2012); these rocks are assigned to petrophysical class 3. However, at the margins (downdip) of the field, L2.0 deposits are subtidal Lower Clear Fork facies. The relatively few data available on these rocks suggest that they are medium-crystalline dolostones assignable to petrophysical class 2.

Lower Clear Fork L2.1

As documented by Ruppel and Harrington (2012), L2.1 consists of a basal section of transgressive to

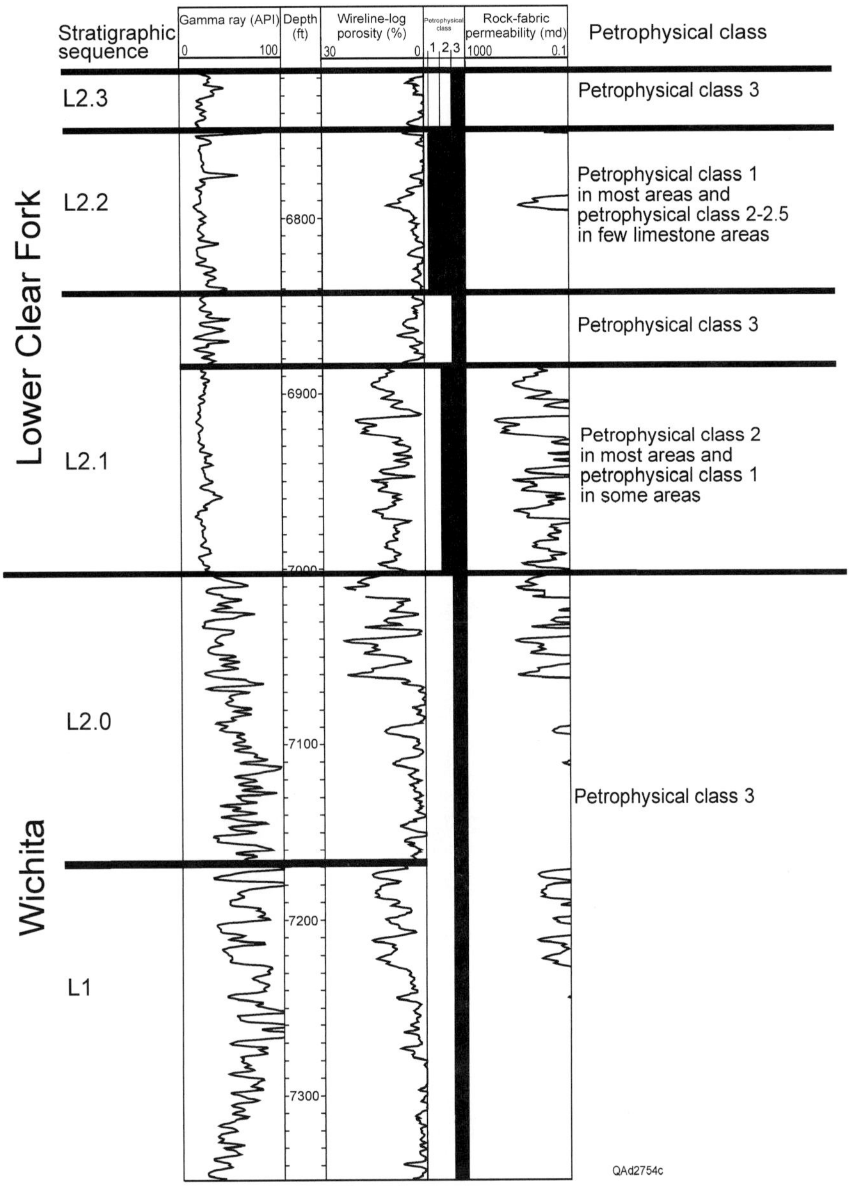

FIGURE 6. Type log showing the nomenclature used in this study and general vertical stacking of petrophysical classes. Wireline logs from FCU 5927 include gamma ray, wireline-log porosity, and rock-fabric-calculated permeability. FCU = Fullerton Clear Fork Unit; GR = gamma ray; API = American Petroleum Institute.

early highstand subtidal platform facies and an upper section of highstand tidal-flat facies. In the lower subtidal interval, cycles shallow upward from lower energy wackestones and mud-dominated packstones to higher energy grain-dominated packstones and grainstones. Typical class 3, mud-dominated rock fabrics in the lower part of the sequence have been converted by diagenesis to class 2 medium-crystalline dolostones similar to the class 2 grain-dominated dolopackstones in the upper part of the sequence (Figure 11). Lime grainstones and grain-dominated packstones are interbedded with dolostones in some areas. Although the limestones possess class 1 and 2 fabrics, respectively, they display class 2 and 3 (mostly class 2) relationships on porosity-permeability crossplots (Figure 11). These limestones behave petrophysically as class 2 and 3 rocks because they contain considerable amounts of moldic (separate-vug) porosity instead of only interparticle porosity. Older core-analysis data suggest that class 1 dolostone fabrics are dominant in L2.1 in two areas of the field (Figure 12). These rocks are probably coarse-crystalline dolostones and anhydritic medium-crystalline dolostones. The effect of patchy anhydrite on petrophysical class will be discussed more fully in the section on HFS L2.2. The distribution of lithology and petrophysical classes in L2.1 is shown in Figure 13.

The upper highstand leg of HFS L2.1 consists of peritidal-deposit fabrics, including mudstones, wackestones, and some mud-dominated packstones, many of which display exposure features, similar to those observed in the Wichita (Ruppel and Harrington, 2012). Dolostones are mud-dominated and fine-crystalline

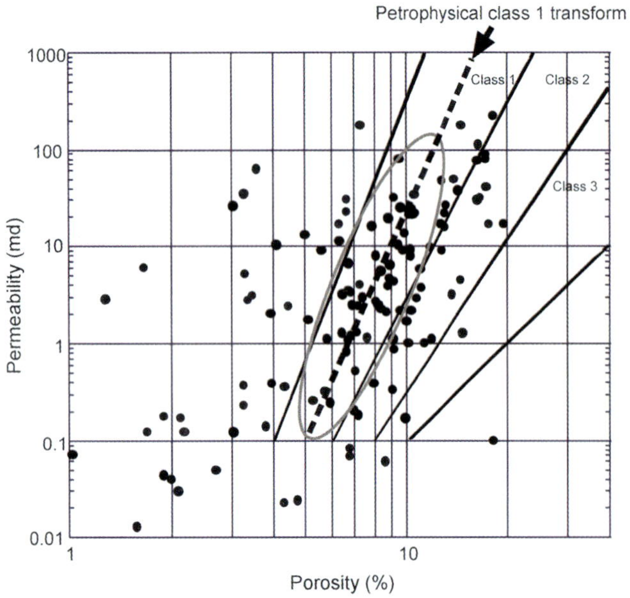

FIGURE 7. Porosity-permeability plot of Abo interval core-analysis data from cores described in this study.

class 3 fabrics (Figure 6); class 3 petrophysical relationships are used accordingly to characterize this unit throughout the field.

Lower Clear Fork L2.2

High-frequency sequence L2.2 is composed dominantly of dolostone, although limestone is locally common in some areas of the field, particularly in the south and in a small area in the northwestern part of the field (Figure 14). Most L2.2 samples studied are class 2 grain-dominated dolopackstones and medium-crystalline, mud-dominated fabrics. However, many samples display class 1 instead of class 2 petrophysical relationships (Figure 15A) because of the presence of 10% or more poikilotopic anhydrite in these medium-crystalline dolostones (Figure 15B). Although counterintuitive, the higher amounts of anhydrite in these rocks actually improve their permeability relative to porosity compared with similar dolostones having less anhydrite because patches of pore-filling anhydrite decrease porosity but have relatively little effect on pore-throat radius. Because permeability is controlled by pore-throat radius, the reduction in porosity caused by anhydrite cementation does not result in a corresponding decrease in permeability. Thus, class 2 rock fabrics containing anhydrite behave like class 1 rock fabrics petrophysically. The anhydrite effect was first observed in the South Wasson Clear Fork reservoir by Lucia and Jennings (2002) and discussed further by Lucia et al. (2004). These rocks are present throughout the field (except where limestone is present) and are best fit by a class 1 transform for estimating permeability and initial water saturation.

Limestone is interbedded with dolostone over most of the southern part of the field and, locally, in the northwest (Figure 14). High-frequency sequence L2.2 rocks in these areas are composed of a mix of low-porosity, medium-crystalline, mud-dominated dolostones (fusulinid wackestones) and high-porosity, grain-dominated limestones (grain-dominated packstones and grainstones). The grainstones mostly contain separate-vug porosity, whereas the grain-dominated packstones tend to contain both separate-vug and interparticle porosity. As with similar rocks in L2.1, these limestones display petrophysical class 2 and 3

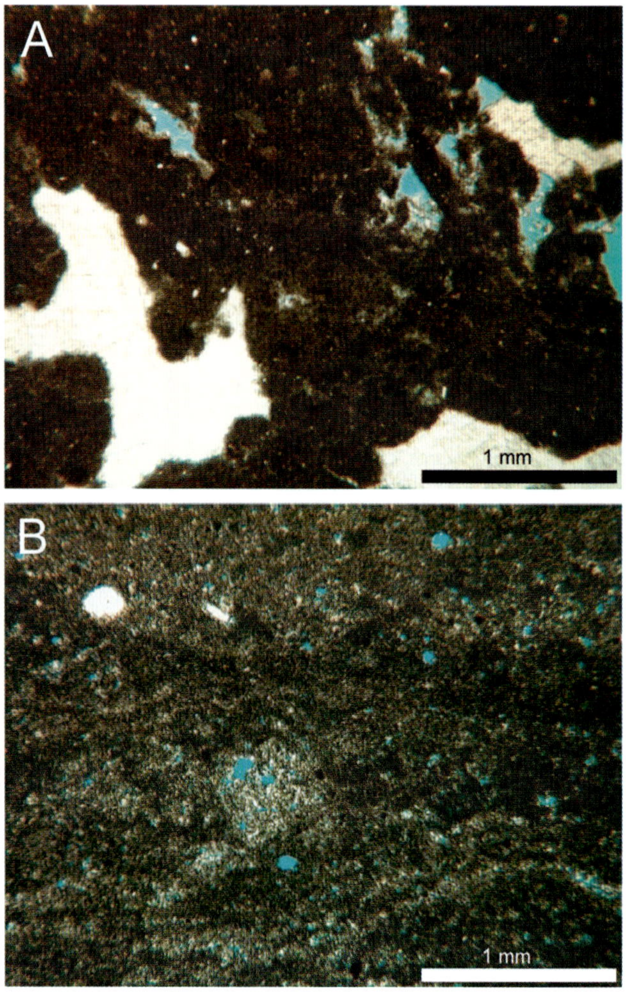

FIGURE 8. Photomicrographs of tidal-flat class 3 fabrics observed in this study. (A) Tidal-flat facies with fenestral pores mostly filled with anhydrite (white): FCU 5927′, 6999′, ϕ = 10.2%, k = 0.08 md. (B) Tidal-flat facies with separate-vug porosity and microporosity: FCU 5927′, 7001′, ϕ = 17.7%, k = 1.26 md. Dolomite crystal size = less than 10 μm. FCU = Fullerton Clear Fork Unit.

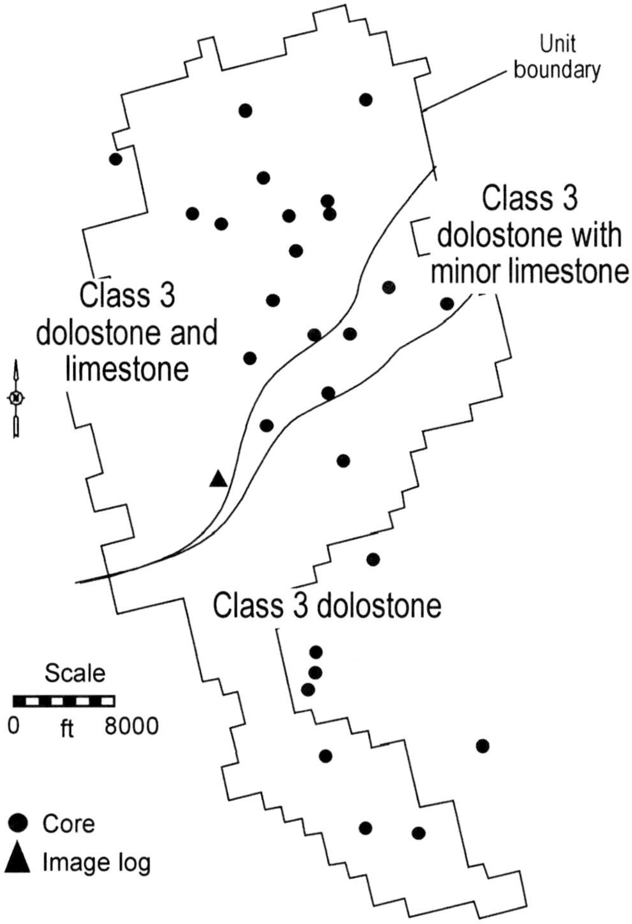

FIGURE 9. Map of Fullerton Unit area showing the distribution of limestone, dolostone, and petrophysical classes in the Wichita Formation.

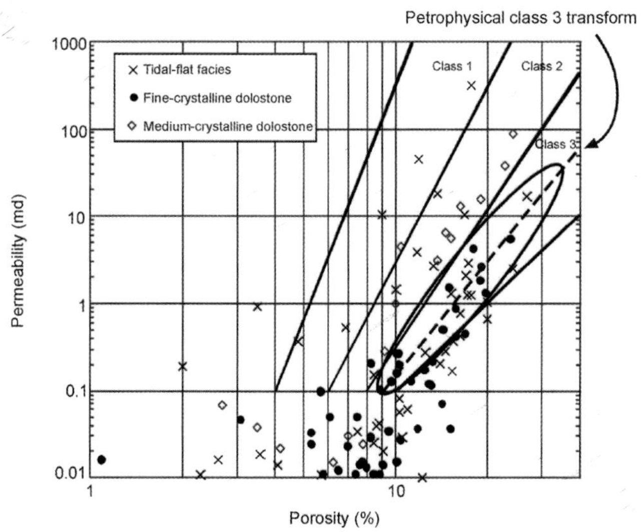

FIGURE 10. Porosity-permeability plot of Wichita facies core-analysis data showing the petrophysical class 3 transform (heavy dashed line) used to calculate permeability. The ellipse shows that most data plot in the class 3 field.

relationships instead of class 1 and 2 because of the presence of moldic pores (Figure 11). As a result, a class 2 transform was used in these areas to estimate permeability and initial water saturation.

Lower Clear Fork L2.3

Tidal-flat-capped, restricted subtidal cycles compose the uppermost Lower Clear Fork HFS (L2.3) throughout most of the field area. Tidal-flat caps are typically fine-crystalline mud-dominated fabrics with

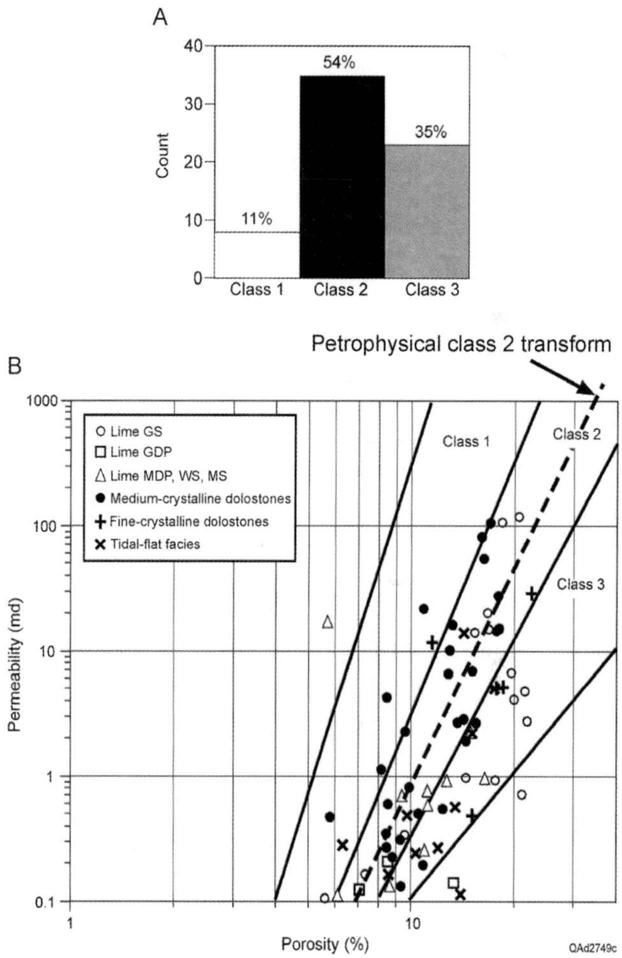

FIGURE 11. Plot and histogram for Lower Clear Fork L2.1 subtidal samples from wells FCU 5927 and FCU 6429. (A) Plot of total core porosity versus permeability showing that class 2 dolostones and higher permeability class 1 moldic limestones both plot in the class 2 field. A petrophysical class 2 transform was used for permeability calculations. (B) Histogram showing that more than 50% of samples plot in the class 2 field. Those that plot in the class 3 field are moldic grain-dominated limestones and class 3 limestones and dolostones. GS = grainstones; GDP = grain-dominated packstones; MDP = mud-dominated packstones; WS = wackestones; MS = mudstones; FCU = Fullerton Clear Fork Unit.

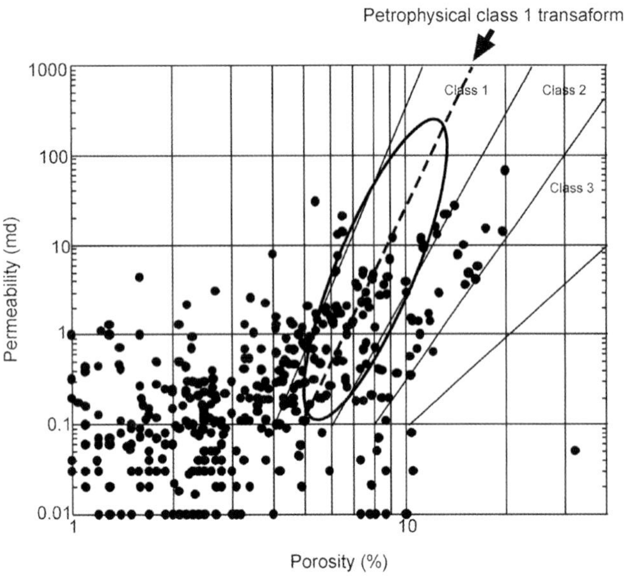

FIGURE 12. **Porosity-permeability plot of core-analysis data from wells with cores displaying dominantly class 1 fabrics in the subtidal part of high-frequency sequence L2.1.**

fenestrae, and the underlying cycle-base subtidal rocks are dominantly fine-crystalline mud-dominated dolostones. Although the cycle-capping, tidal-flat facies are porous and highly correlative across the field, they are very thin and contain little, if any, reservoir permeability. Both the thin tidal-flat caps and the fine-crystalline subtidal dolostones are assignable to petrophysical class 3 over most of the field area. Cycle bases along the downdip eastern margins of the field area appear to contain grain-rich peloidal packstones. However, a detailed study of this facies showed that they are also best characterized by a class 3 transform. Thus, a class 3 transform was used for L2.3 permeability and initial water saturation calculations throughout the field.

DISCUSSION

Limitations Caused by Petrophysical Data Distribution

The detailed rock-fabric data collected in the three cores studied in detail were extrapolated throughout the field using preexisting core data. These new data are necessary to judge the quality and usefulness of the older data. Preexisting core data are variable in their quality, quantity, and distribution throughout the stratigraphic section and must be used judiciously. Figure 16 shows the difference between the new data obtained in this study and the preexisting whole-core data on a porosity-permeability plot. Note that the older data contain a large number of highly variable permeability data in the low-porosity ranges. Such low-porosity, high-permeability data points can indicate problems with core analysis such as incomplete cleaning of samples. Note also that our newer data, which were obtained from carefully cleaned core plugs, contain relatively few data points in this range, confirming the conclusion that some of the older data are flawed. In many cases, we eliminated such old analyses for these reasons.

In addition to basic data-quality issues, previously obtained thin sections were commonly very widely spaced or biased toward a particular facies to yield an adequate description of the vertical variation in rock fabrics and petrophysical classes. More thorough sampling of many of the other cores would have provided more detail on lateral and vertical variations in rock fabric. For example, vertical variations between class 3 and 2 dolostone fabrics are apparent on a scale smaller than the sample spacing in many cores. In the absence of more closely spaced samples, we were forced to use average rock-fabric properties, resulting in a more simplified petrophysical characterization. The impact of both data limitations and rock-fabric averaging is discussed in the following section by stratigraphic unit.

Abo subtidal facies (L1) are poorly known because of the scarcity of wells penetrating the section. Our class 1 rock-fabric assignment for these rocks is based on the porosity-permeability plot of all available core-analysis data from wells with core and confirmed by thin-section study (Figure 7). Areas and intervals of class 2 and 3 fabrics also most likely exist in parts of the Abo that were not sampled, however.

In the Wichita (L1 and L2), most fabrics are class 3, and the entire interval was modeled petrophysically as class 3. However, thin sections and cores indicate local and laterally discontinuous deposits of class 2 dolostone fabrics, particularly in the basal L2 transgressive succession. However, we were unable to define the distribution or continuity of these fabrics and thus were unable to include them in the model. Where present, the permeability of these rocks is probably underestimated because class 3, instead of class 2, transforms were applied. Core studies also show intervals of collapse breccia in the Wichita (Ruppel and Harrington, 2012), suggesting the possibility that a touching-vug pore system composed of fractures and vugs may locally contribute to permeability development. Areas of karsting have not been included in petrophysical modeling because their distribution, except where

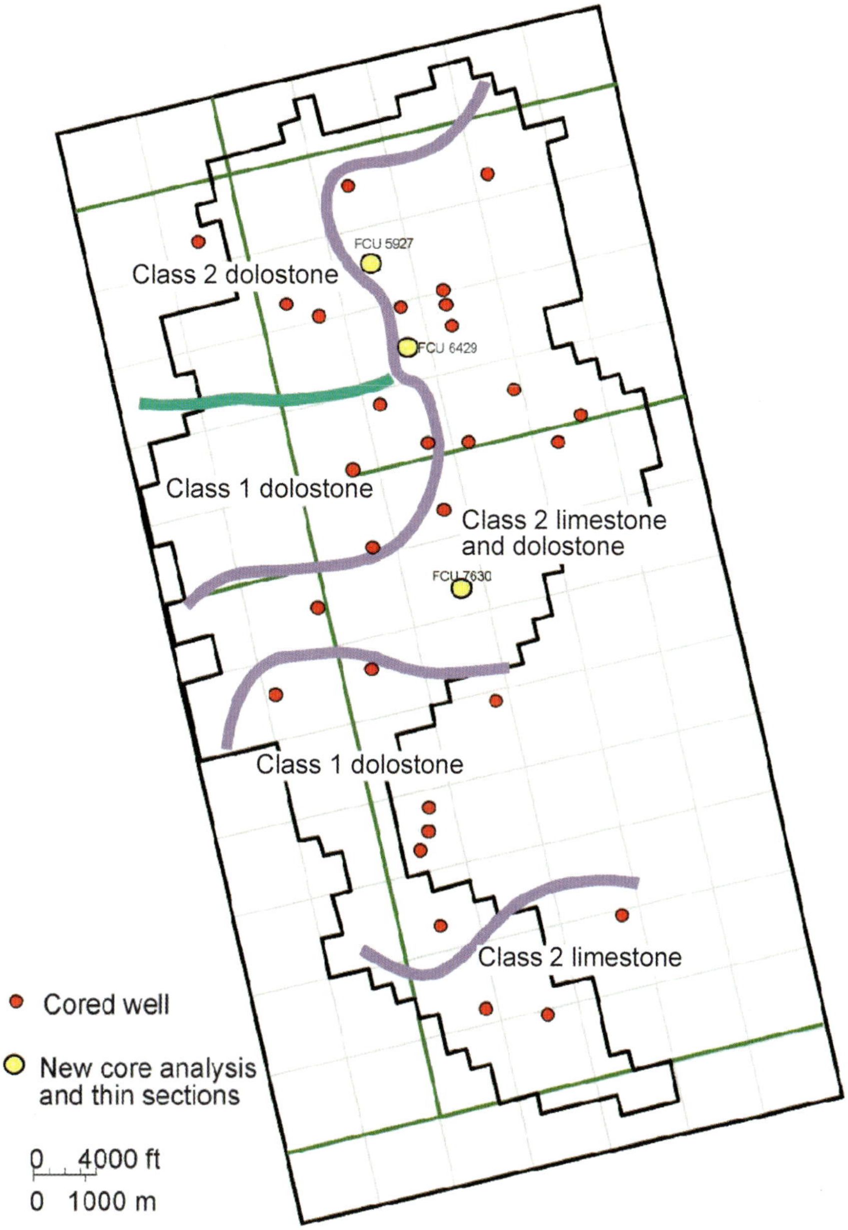

Figure 13. Distributions of petrophysical classes and mineralogy in the subtidal part of high-frequency sequence L2.1.

encountered in cores, is difficult to define and relationships between porosity and permeability (as well as saturation) are not well understood in such systems. As probable evidence of significance of karst development in permeability development and fluid flow, flow simulations required an increase in permeability by a factor of 15 to match injection volumes in the lower Wichita.

In the Lower Clear Fork, major variations in rock fabrics and petrophysical class are mapped in both L2.1 and L2.2. In both sequences, distinct regions of petrophysical class 2 fabrics and class 1 dolostones are apparent. Both sequences also contain areas of interbedded limestone and dolostone having differing rock fabrics. In most cases, limestones are highly porous and grain dominated and contain separate-vug (commonly oomoldic) porosity. Interbedded dolostones are typically lower in porosity, more mud dominated, and have fine to medium dolomite crystals. As described in the Results section, both limestones and dolostones plot in class 2 and 3 fields because of the abundance of moldic (separate vug) pores. Because more data plot in the class 2 field, class 2 was assigned to these areas of interbedded limestone and dolostone. More accurate modeling of the petrophysics of these mixed rock-fabric intervals would require better and more complete log suites than available.

As is apparent from the foregoing, each of the major sequences in the Fullerton reservoir displays a variety of rock fabrics and, accordingly, substantial

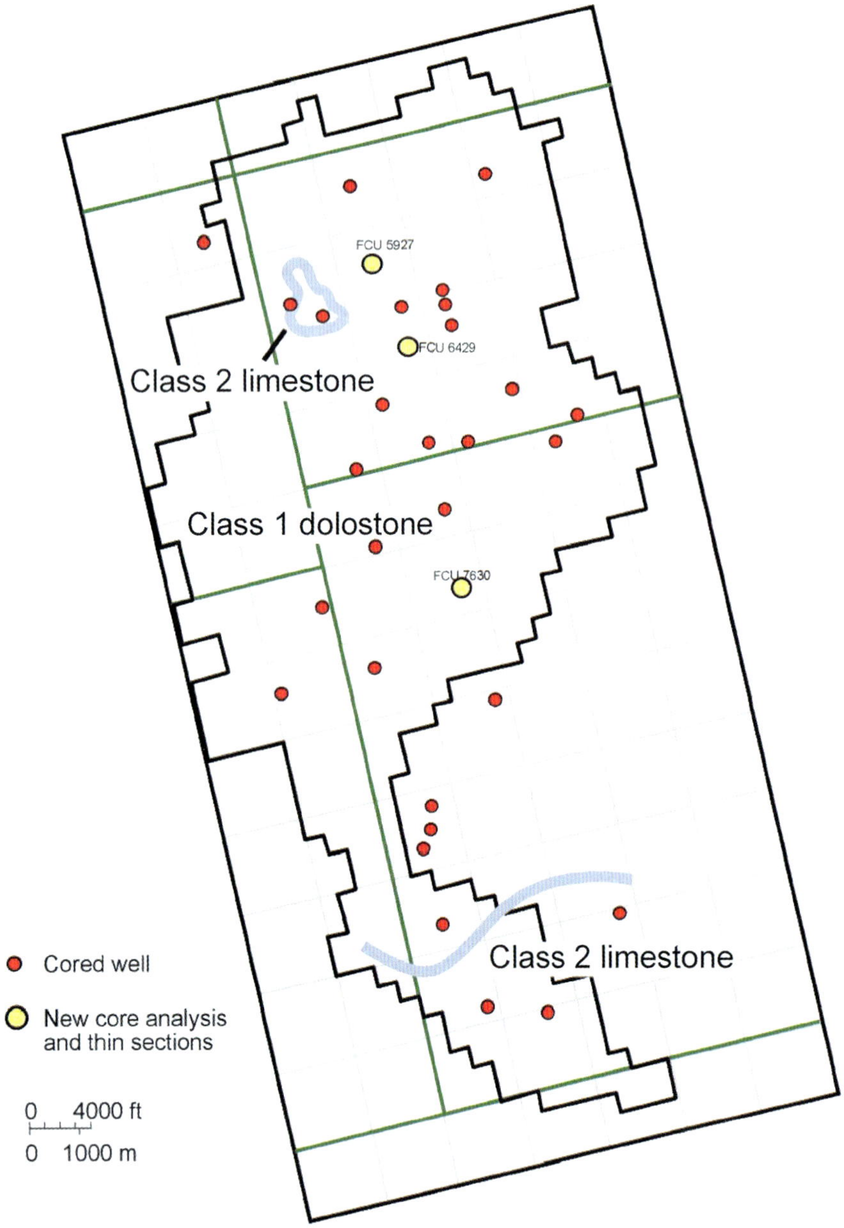

FIGURE 14. Distribution of petrophysical classes and mineralogy in high-frequency sequence L2.2.

stratigraphic and geographic variability in porosity-permeability relationships. For an accurate flow model that includes realistic determinations of variations in permeability and water saturation to be constructed, it is critical that as much rock-fabric-based variability as possible be captured in reservoir models. The core-based rock-fabric characterization methodology previously described facilitates production of cycle-scale maps of the distribution of petrophysical relationships (i.e., petrophysical classes) across the entire field area. This petrophysical quantification of sequence- and cycle-stratigraphic architecture provides the fundamental framework for reservoir modeling and simulation. Lucia and Kane (2012) described the procedures used to calculate permeability and water saturation from the petrophysical-class mapping described here. Wang and Lucia (2012) documented the results of modeling and simulation using these approaches.

CONCLUSIONS

The Clear Fork reservoir at Fullerton field is geologically complex like most carbonate reservoir systems. To capture as much of this complexity as possible, we used a core-based rock-fabric approach to determine and map petrophysical transforms at the cycle scale. Three porosity-permeability transforms, or

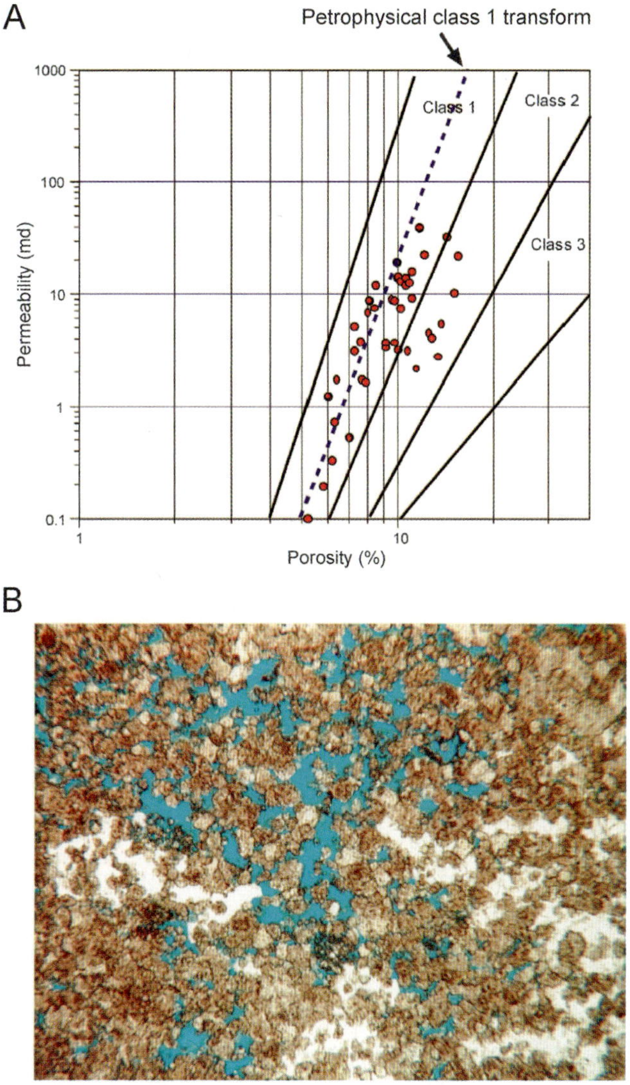

FIGURE 15. (A) Porosity-permeability plot of Lower Clear Fork L2.2 showing that most samples plot in the class 1 field, although they are mostly class 2 medium-crystalline dolostones. (B) Photomicrograph of poikilotopic anhydrite in a class 2 medium-crystalline dolostone having ϕ = 8.1%, k = 9.02 md, and 30% anhydrite. The presence of a large volume of anhydrite shifts this characteristic class 2 rock-fabric sample into the class 1 petrophysical field.

petrophysical classes, were recognized. Fine-crystalline dolostones and mud-dominated limestones characteristic of tidal-flat facies at cycle and sequence tops are petrophysical class 3 rocks. These rocks are dominant throughout the Wichita (sequences L1 and L2.0) and at Lower Clear Fork (L2.1 and L2.2) sequence tops. Lower Clear Fork subtidal facies are more complicated and include rock fabrics assignable to petrophysical classes 1, 2, and 3. Lower Clear Fork dolostones primarily comprise medium- to coarse-crystalline packstones; where they contain large amounts of poikilotopic anhydrite, these facies are class 1 rocks, and where anhydrite is mostly absent, they are class 2 rocks. Lower Clear Fork limestones comprise class 1 and 2 grain-dominated rock fabrics, but the presence of large volumes of separate-vug pores makes them behave petrophysically as class 2 and 3 rocks. This variability clearly

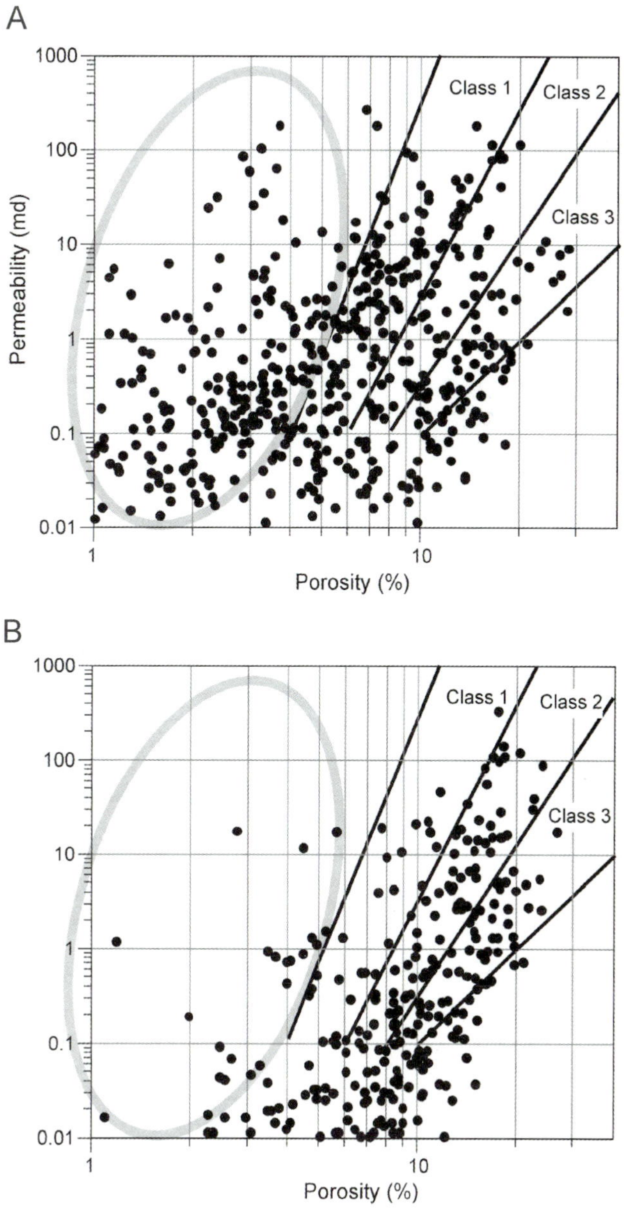

FIGURE 16. Typical core analysis data (A) compared with core analysis data obtained from unbiased sampling and careful cleaning of plugs (B). Note that the large low-porosity-data cloud denoted by the gray ellipse in A is not present in B. The data in A are from FCU 6921 (just south of the flow simulation area); the data in B are the new samples of FCU 5927 and 6429 obtained for this study. FCU = Fullerton Clear Fork Unit.

demonstrates that no single petrophysical transform can adequately define permeability and water saturation in carbonate reservoir systems like Fullerton. Instead, it is critical that multiple transforms be defined and applied to capture both the stratigraphic and geographic variability in petrophysics that is present. Mapping of these complex variations in petrophysical relationships by sequence and cycle provides a strong quantitative basis for developing an accurate framework for flow modeling and simulation. These rock-fabric-based approaches have obvious application to other carbonate reservoir systems worldwide.

ACKNOWLEDGMENTS

This research was funded by ExxonMobil Production Company, The University of Texas System, and the U.S. Department of Energy under contract DE-FC26-01BC15351. ExxonMobil provided a comprehensive data set from the field and access to cores and two-dimensional seismic data. David Smith and Terry Anthony served as invaluable project liaisons. Oxy U.S.A., Inc., also provided logs, data, and access to three-dimensional seismic data and cores. We thank Craig Kemp for his assistance in data transfer. Stephen Ruppel and James Gibeaut provided helpful critical reviews that improved the manuscript immeasurably. Lana Dieterich edited the manuscript. The illustrations were prepared with the assistance of the graphics group at the Bureau of Economic Geology (BEG). The support of this research by the Landmark Graphics Corporation via the Landmark University Grant Program is acknowledged. The publication was authorized by the director of BEG.

APPENDIX 1: METHODOLOGY FOR ROCK-FABRIC DESCRIPTION

Classifying Carbonate Pore Space from Thin Sections

Procedure

What follows outlines the general procedure for describing pore space in petrophysical rock-fabric characterization:

1) **Identify grains.**
   Easily done in most limestones and fine-crystalline dolostones, more difficult but commonly possible in medium-crystalline dolostone, and commonly highly interpretive in large-crystalline dolostone.
2) **Examine intergrain areas and grain sorting.**
   a. Limestones
      If intergrain areas only contain pore space (no mud-size material) or pore-filling minerals (calcite, dolomite, sulfate) and are well sorted, the rock is a grainstone: petrophysical class 1 (Figures 17, 18).
      If intergrain areas contain pore space or pore-filling minerals, as well as mud-size sediment, and if the grains are poorly sorted, the rock is a grain-dominated packstone: petrophysical class 2.
      If intergrain areas only contain mud-size sediment and no intergrain porosity, the rock is a mud-dominated fabric: petrophysical class 3. If the fabric is grain supported, it is classified as a mud-dominated packstone. If it is not grain supported and has more than about 10% grains, it is a wackestone; if it has less than 10% grains, it is a mudstone. Note that the distinction between grain-dominated and mud-dominated packstones is gradational.
   b. Dolostones
      The preceding comments are also applicable to fine-crystalline dolostones and medium-crystalline grain-dominated fabrics without modification. However, dolomite crystal size is important in large-crystalline dolostones and in medium-crystalline, mud-dominated fabrics. All large-crystalline dolostones are petrophysical class 1 without regard to precursor fabric; medium-crystalline mud-dominated dolostones are class 2, not class 3.
3) **Examine areas within grains or crystals.**
   The pore space within grains or crystals is referred to as intraparticle pore space, a type of separate-vug porosity (Figure 19).
   The size of the intraparticle pore space has a large effect on petrophysical properties. If separate-vug porosity is clearly visible (moldic, intrafossil, and others), pores will be oil saturated. However, if the intraparticle pore space is microporosity (faintly visible), pores may be filled with oil and capillary-held water.
4) **Examine pore space significantly larger (>2) than particle (grain or crystal) size.**
   Pores that are significantly larger than particles may be separate vugs or touching vugs, and it is difficult to determine which in thin section (Figure 19). If the large vugs are fabric selective,

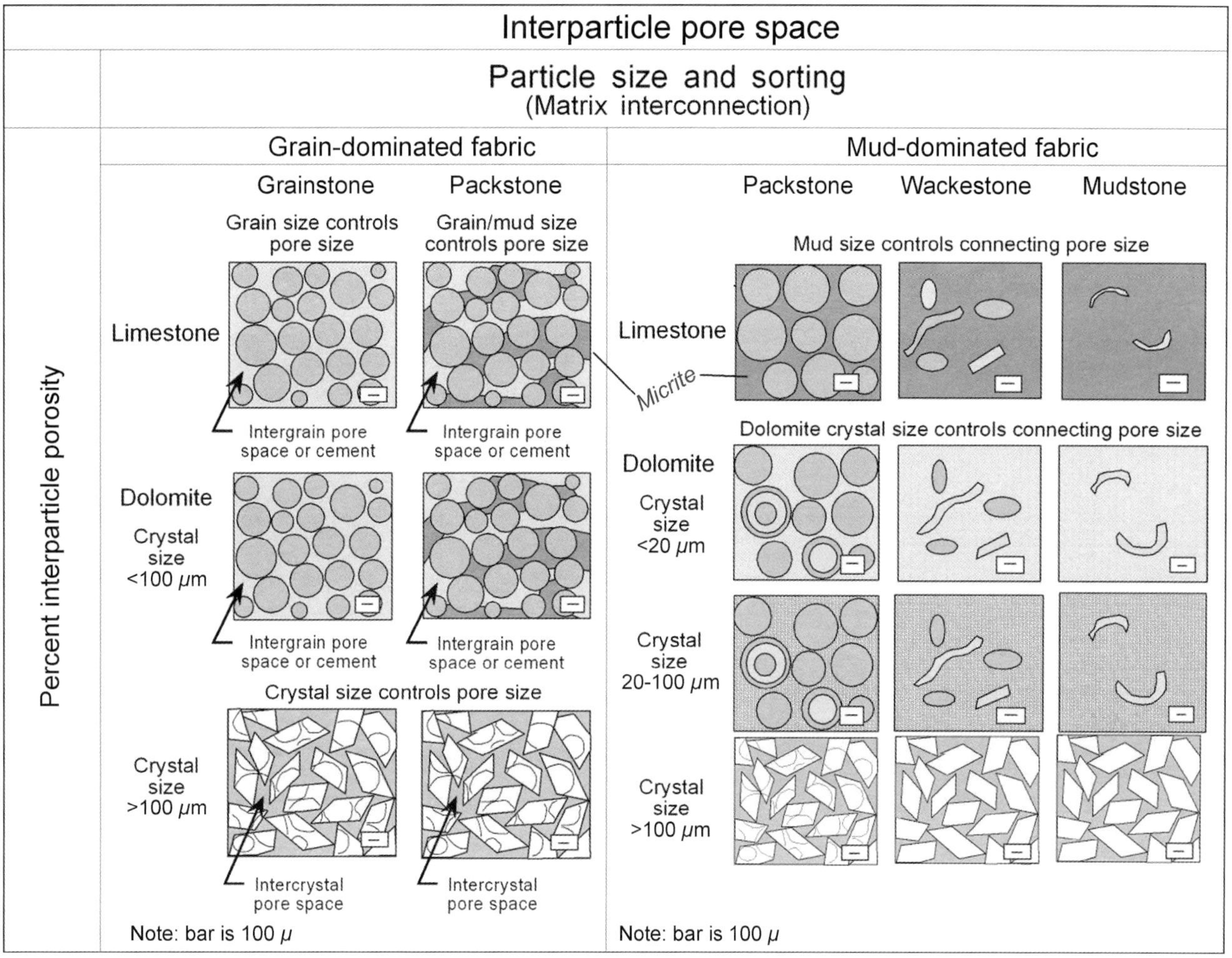

FIGURE 17. **Classification of interparticle pores in carbonate rocks (modified from Lucia, 2007).**

they are probably separate vugs. However, if they are connected by microfractures, they may form a touching-vug pore system. If the large vugs are not fabric selective, they may form an interconnected pore system.

Documentation

The following are the procedures for describing rock-fabric attributes using the attached thin-section description form (Figure 20).

Lithology

Record the crystal size: dolomite crystal size is most important because it controls pore size in mud-dominated dolostones.

Record the type of anhydrite: anhydrite habits are poikilotopic, pore filling, and nodular, which may have important implications as to pore-size distribution. The presence of gypsum is important in log and core analysis.

Texture

Enter the rock-fabric name: grainstone, grain-dominated packstone, mud-dominated packstone, wackestone, or mudstone. Alternatively, show the percentage of mud versus grains graphically. May indicate grain types by name or by symbol.

Petrophysical Fabric

Estimate the interparticle porosity. Interparticle porosity is estimated by subtracting vuggy porosity from core-analysis porosity, if available. May be estimated from thin sections if core analysis is not available.

Record the size of particles, grains, or crystals in micrometers. If the fabric is a grain-dominated packstone, the sizes of particles and interparticle fabric (mud or crystals) are given as a fraction.

Record the separate-vug porosity. Types are grain mold, intrafossil, intragrain microporosity, and others.

Record the presence of touching-vug pores. Types include fracture, vug, and others.

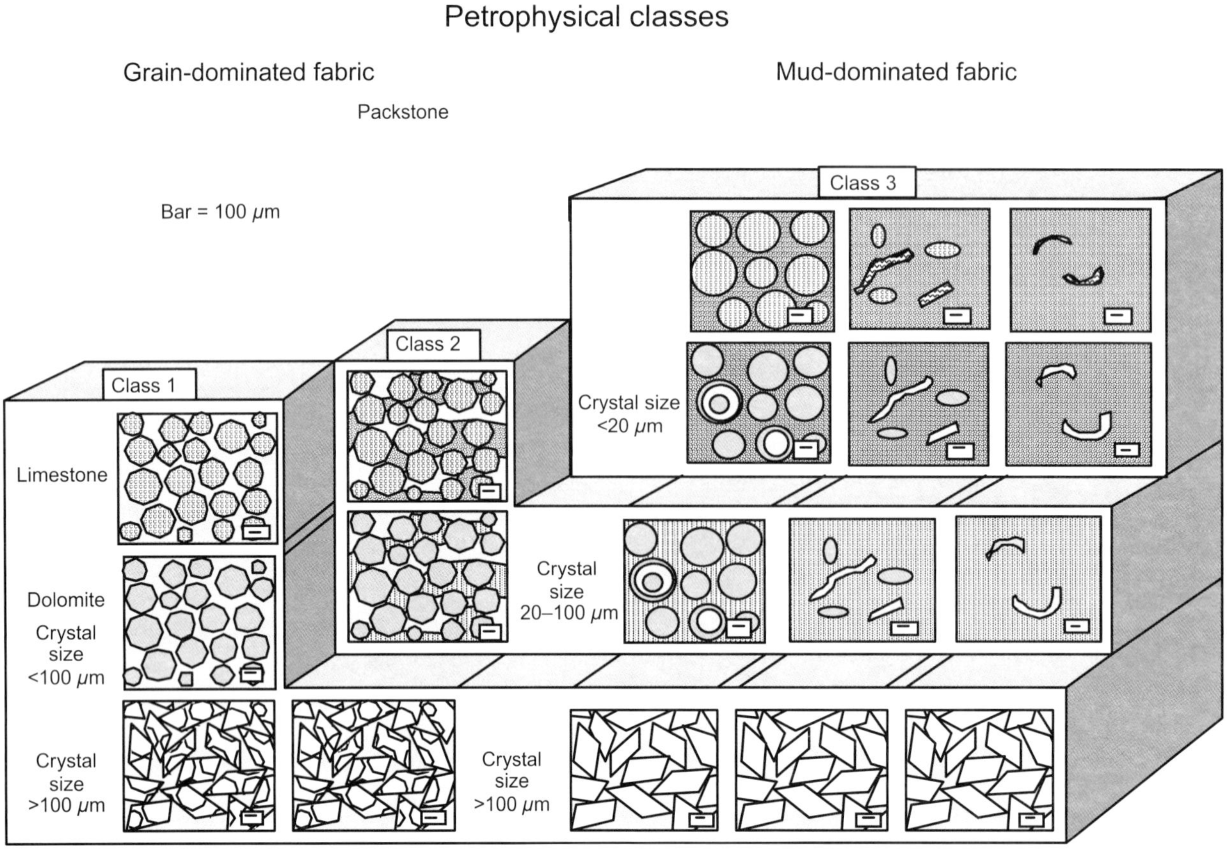

FIGURE 18. Diagram showing the relationship between rock fabrics and petrophysical classes (modified from Lucia, 1995).

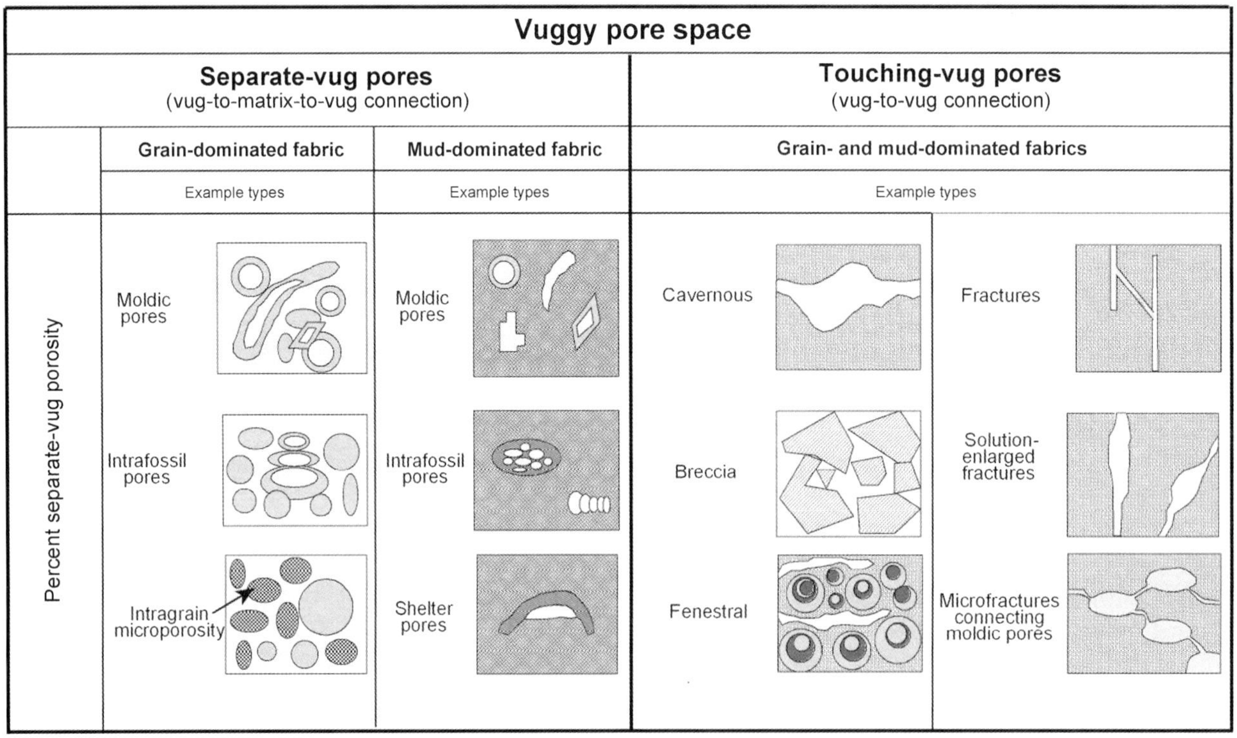

FIGURE 19. Classification of vuggy pore types in carbonate rocks based on vug interconnections (modified from Lucia, 2007).

Depth	Lithology										Texture		Visible pore space				Petrophysical Class	CALC Ippor	Core Analysis		Notes
Sp. No.	Dolomite		Calcite	Sulfate				Quartz		Acc.	Rock fabric	Grain size	Ippor	Svug		Tvug		Cpor-Svug	C_{por}	C_{perm}	
	%	Size (μ)	%	Anhydrite	Type	Gypsum	Type	%	Size		Description	μm	%	%	Type	Type		%	%	md	

FIGURE 20. Worksheet for collecting and recording rock-fabric data (modified from Lucia, 2007). Ippor = interparticle porosity; Svug = separate-vug porosity; Tvug = touching-vug porosity; Acc = accessory; Sp. No. = sample number.

One of three petrophysical classes is normally recorded. However, because classes range from 0.5 to 4, any value within this range is acceptable. See the attachment for the explanation of class range to rock fabrics.

Core Analysis

The data entered here are normally from thin-section samples. Exceptions should be noted.

REFERENCES CITED

Jennings Jr., J. W., and F. J. Lucia, 2001, Predicting permeability from well logs in carbonates with a link to geology for interwell permeability mapping: Society of Petroleum Engineers Annual Technical Conference and Exhibition, New Orleans, September 30–October 3, 2001, SPE Paper 71336, p. 1–16.

Kerans, C., F. J. Lucia, and R. K. Senger, 1994, Integrated characterization of carbonate-ramp reservoirs using Permian San Andres Formation outcrop analogs: AAPG Bulletin, v. 78, no. 2, p. 181–216.

Lucia, F. J., 1995, Rock-fabric/petrophysical classification of carbonate pore space for reservoir characterization: AAPG Bulletin, v. 79, no. 9, p. 1275–1300.

Lucia, F. J., 2000, Petrophysical characterization and distribution of remaining mobile oil: South Cowden Grayburg reservoir, Ector County, Texas: Austin, Texas, The University of Texas at Austin, Bureau of Economic Geology Report of Investigations 260, 54 p.

Lucia, F. J., 2007, Carbonate reservoir characterization: An integrated approach, 2nd ed.: Berlin, Germany, Springer-Verlag, 226 p.

Lucia, F. J., and J. W. Jennings Jr., 2002, Calculation and distribution of petrophysical properties in the South Wasson Clear Fork field, *in* F. J. Lucia, ed., Integrated outcrop and subsurface studies of the interwell environment of carbonate reservoirs: Clear Fork (Leonardian-age) reservoirs, west Texas and New Mexico: Austin, Texas, The University of Texas at Austin, Bureau of Economic Geology, p. 95–142: http://www.osti.gov/dublincore/ecd/servlets/purl/811895-4EHgbz/native/811895.pdf (accessed October 24, 2011).

Lucia, F. J., and J. A. Kane, 2012, Calculation of permeability and initial water saturations from wireline logs in a mature carbonate reservoir, *in* S. C. Ruppel, ed., Anatomy of a giant carbonate reservoir: Fullerton Clear Fork (Lower Permian) field, Permian Basin, Texas: Studies in Geology 63, p. 77–91.

Lucia, F. J., R. H. Jones, and J. W. Jennings, 2004, Poikilotopic anhydrite enhances reservoir quality (abs.): AAPG Annual Convention, v. 13, p. A88.

Lucia, F. J., C. Kerans, and F. P. Wang, 1995, Fluid-flow characterization of dolomitized carbonate-ramp reservoirs: San Andres Formation (Permian) of Seminole field and Algerita Escarpment, Permian Basin, Texas and New Mexico, *in* E. L. Stoudt and P. M. Harris, eds., Hydrocarbon reservoir characterization: Geologic framework and flow unit modeling: SEPM Short Course 34, p. 129–153.

Lucia, F. J., and S. C. Ruppel, 1996, Characterization of diagenetically altered carbonate reservoirs, South Cowden Grayburg reservoir, west Texas: Society of Petroleum Engineers Annual Technical Conference and Exhibition, Denver, Colorado, October 6–9, 1996, SPE Paper 36650, p. 883–893.

Ruppel, S. C., and R. R. Harrington, 2012, Facies and sequence stratigraphy: Critical tools for reservoir framework definition, Fullerton Clear Fork Reservoir, Texas, *in* S. C. Ruppel, ed., Anatomy of a giant carbonate reservoir: Fullerton Clear Fork (Lower Permian) field, Permian Basin, Texas: Studies in Geology 63, p. 5–48.

Ruppel, S. C., and F. J. Lucia, 1996, Diagenetic control of permeability development in a highly cyclic, shallow-water carbonate reservoir: South Cowden Grayburg field, Ector County, Texas, *in* R. L. Martin, ed., Permian Basin oil and gas fields: Keys to success that unlock future reserves: West Texas Geological Society Fall Symposium, p. 7–22.

Wang, F. P., and F. J. Lucia, 2012, Reservoir modeling and simulation of the Fullerton Clear Fork reservoir, Andrews County, Texas, *in* S. C. Ruppel, ed., Anatomy of a giant carbonate reservoir: Fullerton Clear Fork (Lower Permian) field, Permian Basin, Texas: Studies in Geology 63, p. 111–143.

Wang, F. P., F. J. Lucia, and C. Kerans, 1998, Modeling dolomitized carbonate-ramp reservoirs: A case study of the Seminole San Andres Unit: Part I. Petrophysical and geologic characterizations: Geophysics, v. 63, no. 6, p. 1866–1875, doi:10.1190/1.1444479.

3

Kane, J. A., and J. W. Jennings Jr., 2012, Developing a wireline-log database and a basis for determining reservoir porosity, *in* S. C. Ruppel, ed., Anatomy of a giant carbonate reservoir: Fullerton Clear Fork (Lower Permian) field, Permian Basin, Texas: Studies in Geology 63, p. 67–76.

Developing a Wireline-Log Database and a Basis for Determining Reservoir Porosity

Jeffrey A. Kane
Kinder Morgan, Houston, Texas, U.S.A.

James W. Jennings Jr.
Shell, Houston, Texas, U.S.A.

ABSTRACT

We present the procedures used to develop an accurate wireline-log database for the Fullerton Clear Fork field reservoir. We first review the borehole logging history of the field then discuss the assembly and quality checking of the log data and normalization of neutron-porosity logs, and calibration to core porosity. The procedures documented here deviate somewhat from the norm in that the assembly of the database and quality control of data occurred concurrent with data interpretation. Because of the complex nature of the field data, traditional log-normalization methods were not adequate. A method of normalization is described that is a modification of traditional methods, which overcomes the unique problems of the Fullerton Clear Fork reservoir.

INTRODUCTION

In this chapter, we describe the procedures used to assemble, quality check, and interpret wireline logs from Fullerton Clear Fork field, northwestern Andrews County, Texas. As documented by Ruppel and Harrington (2012) and Harrington and Lucia (2012), the reservoir is dominantly composed of dolomite, with an average core porosity of slightly more than 6% and a maximum core porosity of about 33%. Discovered in the 1940s, the field has wireline logs that span six decades of technology.

The assembly and quality checking of the log data were complicated by continued acquisition of new and redigitized log data throughout the study period. The constant influx of data throughout the project required specialized dynamic and iterative updating processes to keep the database current. With such a large field, data normalization is an issue. Several common normalization methods were tried but were found inadequate. The final normalization method used is different from traditional methods in that the mean and standard deviation of the data at each well are estimated from those of nearby wells using a

DOI:10.1306/13341538St633526

weighted moving average. Normalized logs are then shifted to these new local values. Finally, normalized data are calibrated to available core-porosity data. Only porosity was calculated in this study. The procedures used to calculate permeability and water saturation are discussed by Lucia and Kane (2012).

OVERVIEW OF WIRELINE-LOG DATABASE

The Fullerton Clear Fork field contains log suites that span a logging history of more than 60 yr. Because the field was discovered in the mid-1940s, earliest log suites consist of electric logs (or e-logs) common to that period. Moreover, because the field has had several periods of development throughout its history, log suites vary from these original e-logs to modern, more sophisticated, and varied suites available today. This long time span of data collection created numerous difficulties for log quality control (QC) and analysis.

Log suites in Fullerton Clear Fork field can be divided into four general groups. The first group is composed of those logs associated with initial field development in the mid- to late 1940s and includes early e-logs. The second group, characteristic of 1950s logs, differs from the first group in the addition of a gamma-ray neutron log. The third group essentially consists of a sidewall neutron log and a modern resistivity log, either a dual laterolog or a dual induction. This log suite was common throughout the 1960s and was the first suite to contain a log calibrated in porosity units. During the 1970s, the use of the compensated neutron log was introduced, commonly without resistivity logs. The use of this suite has continued to the present.

The oldest e-logs are composed commonly of three tools: a long lateral of varying spacing length; a short normal, again of varying spacing; and a limestone lateral curve. These long laterals had spacings of 13, 18, and 24 ft (4, 5, and 7 m) generally; however, some other lengths were used commonly. The short normal typically had 10 in. (25 cm) spacing; however, 12-, 16-, and 18-in. (30, 41, and 46 cm) spacings were also used. Limestone e-logs were essentially all 32 in. (81-cm) laterals, along with a few 34 in. (86 cm) laterals. For the most part, these log suites were not used for evaluation in the field, except in specialized circumstances in which attempts to evaluate original oil saturation profiles were made.

The second group of logs, which includes those that contain some form of gamma-ray neutron log, is composed of wells that were logged during the 1950s to the early 1960s. Unfortunately, these logs considerably vary in tool design, measurement type, and overall quality. Numerous problems properly defining the logging environment and a lack of core information for calibrating these data to porosity precluded using this group of wells in field analysis.

The third group of logs, as previously stated, is composed of sidewall neutron logs (SNP) typically run with a dual laterolog. This is the oldest suite of logs with calibrated porosity curves and core sufficient to provide a calibration to core-measured porosities. This set of data is composed of logs from 320 wells, 316 of which have data acceptable for normalization. The locations of these 316 wells are displayed in Figure 1. The actual process of porosity normalization is discussed in the porosity log normalization section of this chapter.

The fourth group of logs essentially includes all modern-day logs. The dominant porosity log in this group is the compensated neutron log (CNL). This data set is made up of logs from 471 wells, 437 of which were of sufficient quality to be used for normalization (Figure 2).

Twenty wells contain logs common to groups 3 and 4. Because of this overlap, the total number of wells used for the quantitative part of the full-field study is 733. The locations of these wells are shown in Figure 3. As can be seen from this figure, well coverage is good over most of the field; however, the western edge and some areas of the southern part lack coverage sufficient for detailed characterization.

DATABASE ASSEMBLY PROCEDURES

In order to build and maintain the log database, we selected log analysis software based on its ability to handle large numbers of logs and, more importantly, process large numbers of logs in batches relatively quickly. The ability to efficiently run batch calculations greatly facilitated the dynamic updating of the log database.

QUALITY CONTROL

The initial Fullerton field database included about 1150 wells. Because of concerns over the quality of the original database, the field operator, ExxonMobil, began a complete redigitization of all logs in the field soon after the reservoir characterization study began.

FIGURE 1. Map showing 316 wells with sidewall neutron logs used to calculate porosity for Fullerton Clear Fork field.

Ultimately, 1206 wells were redigitized. The digitization contractor, IHS Energy, provided all initial QC with respect to the actual digitization project. The data were then sent to ExxonMobil, which forwarded the data to the Bureau of Economic Geology for inclusion in the characterization project. During the initial phase of the project, detailed QC was performed on logs from about 275 wells after digitization. These wells included examples of all log-suite types. This stage of QC provided feedback on needed modifications to the digitizing process and helped to resolve issues and problems early in the process. Throughout the project, all well-log data were reviewed before loading for completeness of digitization; detailed QC was performed randomly.

For QC of digitized log data to be maintained properly, copies of original paper logs are critical. Visual verification of the quality of digitization, verification of proper log-header transcription, and review of paper logs for wellbore-environment specifics affecting

FIGURE 2. Map showing 437 wells with compensated neutron logs used to calculate porosity for Fullerton Clear Fork field.

proper log analysis can be done only with access to copies of original paper logs. For example, during the review of gamma-ray neutron logs, it became apparent that at least one half had been run during well-deepening events. This meant that these logs had both open- and cased-hole components. Such logs need to be separated into open- and cased-hole logs and normalized independently. Without paper logs, this situation would not have been recognized. Further investigation of these gamma-ray logs revealed a broad spectrum of gamma-ray neutron tool types and a lack of cores for calibration, making normalization of these logs unreliable. Thus, these logs were omitted from the project.

Well logs for a given well in an update could have several files associated with it, and each of these files would contain log data at different levels of correction. Not all log types were included in the correction process or in all levels of the correction process, so the loading of all files was necessary to capture all the data. Because of differences in data quality by file, the data had to be loaded in a specific order to ensure that the most recent, most completely edited curve was primary. For a few wells, this process could be

FIGURE 3. Map showing all 733 wells used in porosity calculation for Fullerton Clear Fork field.

done interactively, but the large number of wells in this project made interactive loading unrealistic, requiring that an automated or semiautomated process be developed instead. Each file was ordered on the basis of the level of correction and date received. Once this ordering was done, all data were copied into a file from the oldest unedited data, to the oldest edited data, to the most recent unedited data, and finally, to the most recent edited data. At the conclusion of this process, a single file existed that contained all log data for a given well in a single table. The process culminated in 17,200 logs in the final update. Each time that new data were acquired, the entire database was rebuilt completely to guarantee a complete data set.

Each time that the database was rebuilt, available log curves were reviewed. The digitization provider was commonly inconsistent in the naming conventions used, so curve mnemonics referring to the same data type varied throughout the digitization process. A total of 525 different log mnemonics were found in this review at the end of the project. Rectifying the differences was beyond the scope of this project and was performed only on an as-needed basis.

POROSITY-LOG NORMALIZATION

The first step in determining the basic petrophysical parameters in this study was normalization of usable data. For this study, because the primary goal was the accurate assessment of porosity, the normalization process exclusively focused on porosity logs, specifically, CNL and SNP. Two different normalization methods were used. A traditional normalization method was used in the early stages in a small area of the field selected for detailed study, modeling, and simulation. In later full-field studies, this method of normalization was not adequate without modification.

Traditional normalization methods fall into two general categories. The first category of normalization methods requires stratigraphic intervals of constant properties over the area to be considered. For instance, in normalizing a porosity log, a low-porosity interval, such as an anhydrite, and a high-porosity interval, such as shale, would be determined in each well, and values of each log (CNL and SNP) for these intervals would be averaged for the area of interest. Each well log would then be shifted so that its average value across those intervals would be the same as the global average for that log in the respective stratigraphic interval. However, the Fullerton field has no such intervals that can be used, making this approach inappropriate.

The second category of normalization methods commonly used requires an assumption that a distribution of some statistical parameter(s) (i.e., minimum and maximum values or the average and standard deviation of the porosity) is constant over some thick interval. The assumption of constant average and standard deviation of porosity was used in the initial study area. The approach was reasonable, given the small (~2 mi^2 [~3 km^2]) size of the initial study area. When this approach was applied to the full-field normalization of the neutron-porosity logs, it was found that these parameters are not constant across the entire field (~45 mi^2 [~117 km^2]), so a variation of this method was used.

The method used in cases with areally varying statistics follows the work of Jennings et al. (2002). First, CNL or SNP porosity values were vertically averaged separately over genetic intervals. In Fullerton, stratigraphic cycles were the genetic intervals used. Using several intervals, a vertical average CNL or SNP porosity value is determined for all cycles under consideration by well. The vertical standard deviations of these cycle averages were also computed by well. In the Fullerton study, 19 such cycles were used, including cycles for the Lower Clear Fork to upper Wichita interval.

This computation by using genetic intervals minimizes random errors, such as measurement noise, while preserving geologically relevant information. This phase of the process does not correct for systematic errors, such as those generated by the incorrect calibration of the logging tool. Such systematic errors are uniform over a single log at a single well location. As such, the error will be reflected in the mean and standard deviation computed for a given well. Combining the means and standard deviations over some wells, as would be done in computing a moving average, suppresses the systematic error associated with a single well.

Areally averaging the vertically averaged porosities and standard deviations of the averaged porosities over a subset area will provide a way of estimating a slowly laterally varying average and standard deviation of the CNL or SNP porosity. This use of a moving average can be thought of as a low-pass filter removing high-frequency local variability to enhance larger scale trends. The mean and standard deviation predicted via this method are now taken to be correct values at the well. To correct the neutron-porosity log curve to this new value for mean and standard deviation, each data value (CNL or SNP) is adjusted with the equation:

$$\Phi_{\mathrm{corr}} = (\Phi_{\mathrm{orig}} - \mu_{\mathrm{orig}})\frac{\sigma_{\mathrm{pred}}}{\sigma_{\mathrm{orig}}} + \mu_{\mathrm{pred}}, \quad (1)$$

where Φ_{corr} is the corrected or normalized CNL or SNP data value, σ_{pred} is the new or predicted standard deviation, σ_{orig} is the standard deviation calculated from the original log data, Φ_{orig} is the original CNL or SNP data value, μ_{pred} is the new or predicted mean, and μ_{orig} is the mean calculated from the original data.

The CNL and SNP logs were handled separately because their lithology response in dolomite is substantially different. Typically, CNL and SNP logs are run on the assumption that the matrix being logged is limestone. If the actual matrix is not limestone, then the calculated porosity will be in error. For SNP logs, this porosity difference is about two porosity units too high in dolomite. The CNL variations in dolomite are not constant; however, in the porosity range encountered in Fullerton reservoir rock, the difference is about six porosity units too high for dolomite. Because of this difference in dolomite response,

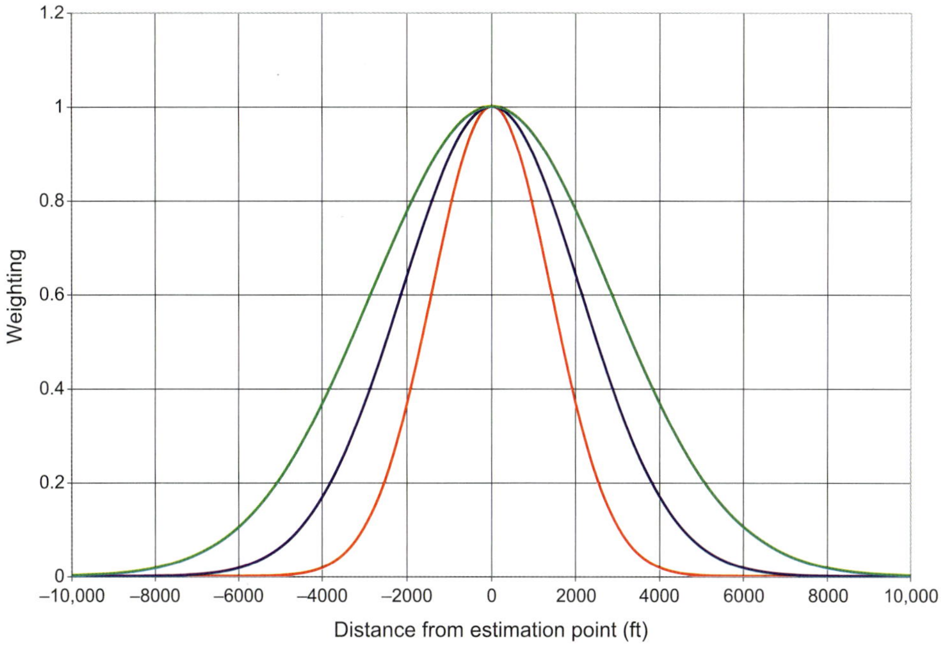

FIGURE 4. Example of a Gaussian weighting function using various scale factors. The red curve represents a scale factor of 2000 ft (610 m); blue, 3000 ft (914 m); and green, 4000 ft (1219 m). A scale factor of 3000 ft (914 m) was used for porosity-log normalization in the Fullerton field study.

the two different tools were handled separately and combined into a single data set after calibration to available core porosity.

MOVING-AVERAGE COMPUTATION

Many different weighting functions can be used to compute a moving average. For this study, a Gaussian weighting function was used:

$$w_i = e^{-\left(\left(\frac{x-x_i}{s}\right)^2+\left(\frac{y-y_i}{s}\right)^2\right)}, \qquad (2)$$

where w_i is the weight computed at the location x_i, y_i; x and y are the x and y locations of the well, respectively; and s is the scale factor.

This function decays toward zero as the distance increases from the point where estimation is occurring but never goes to zero. All data are always used, although the weight of each point decreases with increasing distance from the estimation point (Figure 4). The fundamental parameter controlling the Gaussian weighting function is the scale factor, which influences the decay rate of the weighting value computed by the Gaussian function with respect to distance. An optimal scale factor would suppress local variation without unduly suppressing the larger scale variations. In Fullerton field, this parameter is chosen by inspection. Several maps were made with varying scale factors (Figure 5). After inspection, a scale factor of 3000 ft (914 m) was used because it visually suppressed local variations in porosity without oversmoothing.

CORE CALIBRATION

Once the data are normalized, they must be corrected to represent porosity, commonly done by calibration to available whole core data. All available core data were used in this calibration process. For each curve used, two different calibration lines were derived: one for dolomite, and one for limestone. The plots showing the calibration lines for both compensated neutron logs and sidewall neutron logs are provided in Figures 6–9.

The method used to determine calibration lines differed slightly between limestone and dolomite. Because dolomite is most prevalent throughout the reservoir interval, the approach for generating dolomite calibration lines was consistent with the averaging scheme used to determine normalization estimators. That is, core data were averaged over the same stratigraphic cycles as those for normalized neutron-porosity logs. The goal was to eliminate, as much as possible, the difference in vertical resolution between the two measurements. In addition, vertical averaging reduces the random noise associated with the measurements.

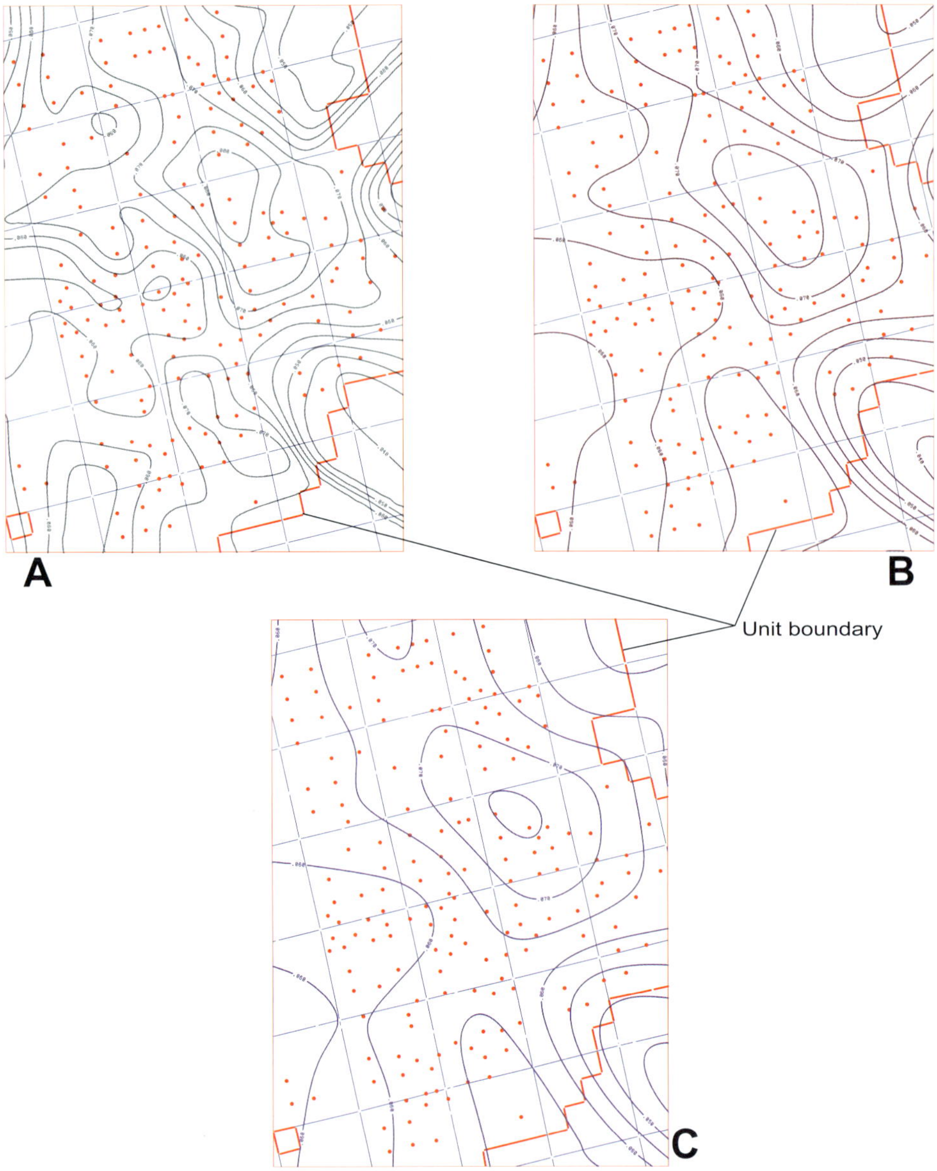

FIGURE 5. Examples of average sidewall-neutron-log porosity areally averaged using different scale factors. Map A was generated using a 2000 ft (610 m) scale factor; map B, a 3000 ft (914 m) scale factor; and map C, a 4000 ft (1219 m) scale factor. The red circles represent well control. Map A still contains "bull's eyes," contours controlled by a single data point. Map C shows a lack of local variability, completely dominated by larger scale trends. Map B represents a reasonable compromise between local variability and larger scale trends. A scale factor of 3000 ft (914 m; map B) was used for porosity normalization in Fullerton Clear Fork field.

The limestone calibration had to be handled somewhat differently. Because limestones are not as pervasive as dolomites in this reservoir, the aforementioned method would not work. The averaging scheme used to generate dolomite porosity calibrations suppresses the limestone response of the logs because of the relative percentage of limestone being much less than that of dolomite in this reservoir. Limestone calibration was therefore computed by comparing neutron-porosity data with core data in the limestone intervals. Which data points corresponded to limestone was determined by selective filtering of the photoelectric factor (PEF) curve. Because calcite has a PEF value of 5 and dolomite has a PEF value of 3, if a limit of accepting only a PEF value greater than 4 is set, limestones would be the prevalent lithology used to create the calibration line in this case. The porosity calibration lines for the four different groups are:

Compensated neutron – dolomite

$$\Phi_{\text{core}} = -0.0231 + 0.82296 \times \Phi_{\text{NPHInorm}}$$

Compensated neutron – limestone

$$\Phi_{\text{core}} = -0.0358 + 1.1709 \times \Phi_{\text{NPHInorm}}$$

Sidewall neutron – dolomite

$$\Phi_{\text{core}} = -0.0178 + 1.0781 \times \Phi_{\text{SNPnorm}}$$

Sidewall neutron – limestone

$$\Phi_{\text{core}} = -0.0238 + 1.248 \times \Phi_{\text{SNPnorm}},$$

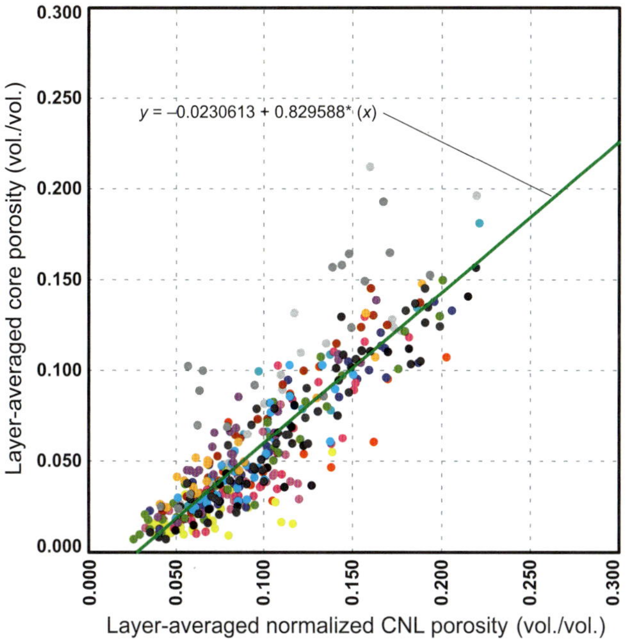

FIGURE 6. Plot of layer-averaged core porosity (*y* axis) as a function of layer-averaged normalized compensated neutron porosity (*x* axis) showing the calibration line for dolomites. Each point represents data that have been vertically averaged over a single cycle. Like colors represent data from the same well. The calibration is based on 21 wells. CNL = compensated neutron log.

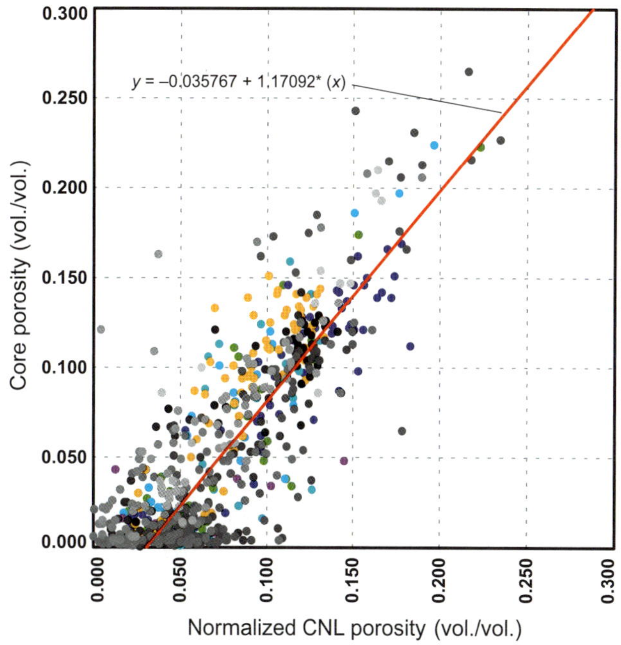

FIGURE 8. Plot of core porosity (*y* axis) as a function of normalized compensated neutron porosity (*x* axis) showing the calibration line for limestone. Each point represents an individual core sample. Only data where the photoelectric factor was greater than 4 were used in generating the calibration. Like colors represent data from the same well. The calibration is based on 21 wells. CNL = compensated neutron log.

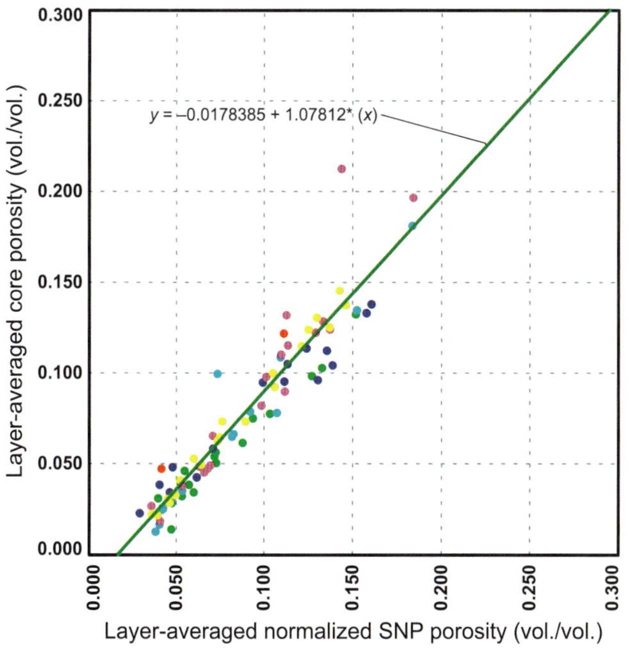

FIGURE 7. Plot of layer-averaged core porosity (*y* axis) as a function of layer-averaged normalized sidewall neutron porosity (SNP) (*x* axis) showing the calibration line for dolomites. Each point represents data that have been vertically averaged over a single cycle. Like colors represent data from the same well. The calibration is based on six wells.

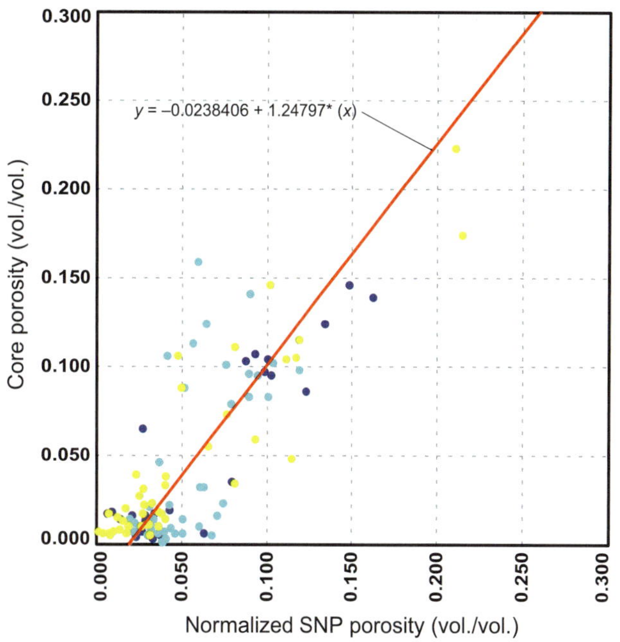

FIGURE 9. Plot of core porosity (*y* axis) as a function of normalized sidewall neutron porosity (SNP) (*x* axis) showing the calibration line for limestone. Each point represents an individual core sample. Only data where the photoelectric factor was greater than 4 were used in generating the calibration line. Like colors represent data from the same well. The calibration is based on five wells.

where Φ_{core} is the core-calibrated log porosity, $\Phi_{NPHInorm}$ is the normalized compensated neutron porosity, and $\Phi_{SNPnorm}$ is the normalized sidewall neutron porosity.

CONCLUSIONS

Porosity evaluation in large fields requires detailed attention to data QC, but the level of QC needed in a digital database can be acquired only through access to the original paper logs used to generate the digital data. Transcription and digitization errors can occur and, in some cases, wellbore information present on the paper logs is never captured.

Because of the timing of log digitization, methods had to be developed for updating the database as new and reedited log data were received to guarantee that the logs being used were the most current, most edited versions. In the case of Fullerton Clear Fork, the sheer volume of data necessitated automating many of the processes.

Proper normalization is necessary to allow correct calibration to core porosity fieldwide. Traditional porosity-log-normalization methods failed to provide acceptable results. A novel normalization method was therefore applied that overcomes the shortcomings of the other methods tried. This method, which involves both vertical averaging of log data by genetic interval and areal moving averages to extract trends in the local mean and standard deviation of target log data, is general in nature and should have a wide applicability. The dual averaging, vertically and areally, has the effect of minimizing the impact of both random errors such as measurement noise and systematic errors associated with logging tool miscalibration. The normalization method outlined has advantages over other normalization methods in that any reasonably thick interval can be used to normalize the log, even the reservoir interval itself.

ACKNOWLEDGMENTS

We thank Florence Bonnaffé and Stephen C. Ruppel for their constructive review comments on the manuscript. Lana Dieterich edited the manuscript. The illustrations were prepared with the assistance of the graphics group at Bureau of Economic Geology (BEG). The support of this research by the Landmark Graphics Corporation via the Landmark University Grant Program is acknowledged. The publication was authorized by the director of BEG.

REFERENCES CITED

Jennings Jr., J. W., F. J. Lucia, and S. C. Ruppel, 2002, 3-D modeling of stratigraphically controlled petrophysical variability in South Wasson Clear Fork reservoir: Society of Petroleum Engineers Annual Technical Conference and Exhibition, San Antonio, Texas, September 29–October 2, 2002, SPE Paper 77592, 15 p.

Harrington, R. R., and F. J. Lucia, 2012, Integration of rock fabrics and stratigraphy for petrophysical quantification of reservoir framework, *in* S. C. Ruppel, ed., Anatomy of a giant carbonate reservoir: Fullerton Clear Fork (Lower Permian) field, Permian Basin, Texas: Studies in Geology 63, p. 49–66.

Lucia, F. J., and J. A. Kane, 2012, Calculation of permeability and initial water saturations from wireline logs in a mature carbonate reservoir, *in* S. C. Ruppel, ed., Anatomy of a giant carbonate reservoir: Fullerton Clear Fork (Lower Permian) field, Permian Basin, Texas: Studies in Geology 63, p. 77–91.

Ruppel, S. C., and R. R. Harrington, 2012, Facies and sequence stratigraphy: Critical tools for reservoir framework definition, Fullerton Clear Fork Reservoir, Texas, *in* S. C. Ruppel, ed., Anatomy of a giant carbonate reservoir: Fullerton Clear Fork (Lower Permian) field, Permian Basin, Texas: Studies in Geology 63, p. 5–48.

4

Lucia, F. J., and J. A. Kane, 2012, Calculation of permeability and initial water saturations from wireline logs in a mature carbonate reservoir, *in* S. C. Ruppel, ed., Anatomy of a giant carbonate reservoir: Fullerton Clear Fork (Lower Permian) field, Permian Basin, Texas: Studies in Geology 63, p. 77–91.

Calculation of Permeability and Initial Water Saturations from Wireline Logs in a Mature Carbonate Reservoir

F. Jerry Lucia and Jeffrey A. Kane[1]

Bureau of Economic Geology, The University of Texas at Austin, Austin, Texas, U.S.A.

ABSTRACT

Estimating permeability from wireline logs has been difficult historically because of the large petrophysical heterogeneity typical of carbonate rocks. A simple relationship between porosity and permeability has not been observed, only between porosity, permeability, and pore size. In Clear Fork carbonates at Fullerton field, pore size is related to rock fabrics. Permeability was calculated for each well using porosity from wireline logs and rock-fabric information from stratigraphic relationships. Permeability profiles, calculated using a global transform, compare well with core permeability values. An exception is in the lower section of the reservoir (Wichita Formation), where karst fabrics are present, suggesting a touching-vug pore system.

Initial water saturation is required for estimating original oil in place and for flow simulation studies. Because most of the original wells in this field had very old log suites and the new wells were drilled after the advent of waterflooding, calculation of initial water saturation from wireline logs for most of these wells was not feasible. In this study, initial water saturations were calculated using capillary-pressure models generated for each petrophysical rock-fabric class. A generic rock-fabric model for class 1 fabrics was used, and new rock-fabric models were developed for class 2 and class 3 fabrics using the Thomeer approach. Water saturations calculated from capillary-pressure models show good agreement to

[1] *Present address:* KinderMorgan, Houston, Texas, U.S.A.

DOI:10.1306/13341539St633155

saturations calculated using the Archie equation from wireline logs in intervals thought to be unflooded. The success of the rock-fabric-based method in calculating permeability and saturation in the Fullerton Clear Fork reservoir illustrates how valuable this technique can be in defining the petrophysical properties in carbonate reservoirs.

INTRODUCTION

Estimating permeability from wireline logs has been difficult historically because of the large petrophysical heterogeneity typical of carbonate rocks. Permeability is a function of pore-size distribution and not of simple porosity. A simple relationship between porosity and permeability has not been observed, only between porosity, permeability, and pore size. Sonic, resistivity, and nuclear magnetic resonance tools have been used successfully in some reservoirs for characterizing pore size for use in porosity-permeability transforms. The Fullerton Clear Fork field, Andrews County, west Texas, however, only has gamma-ray and neutron logs available in most of the wells, and pore size cannot be obtained from these logs. In this study, pore size is related to rock fabrics. Permeability is calculated from wireline logs using porosity from wireline logs and rock-fabric information derived from stratigraphic relationships.

Initial water saturation profiles were calculated for each well using a capillary-pressure model. The normal method for calculating water saturation is through the use of porosity and resistivity logs. However, most wells in this field do not have resistivity logs, and saturation models built from capillary-pressure data are used instead. In this study, saturation is related to porosity, rock fabric, and reservoir height using new capillary-pressure data. Similar to permeability calculations, rock-fabric information is obtained from stratigraphic relationships.

CALCULATING PERMEABILITY FROM WIRELINE LOGS

Permeability is a function of pore-size distribution and porosity. In this study, rock-fabric descriptions are used to estimate pore-size distribution, and permeability is calculated using the global transform (equation 1). The global transform was derived from a database composed of core-analysis data and rock-fabric descriptions (Lucia, 1999; Lucia et al., 2001; Jennings and Lucia, 2003). The equation is useful for calculating matrix permeability when petrophysical class (or rock-fabric number) and interparticle porosity are known.

$$\mathrm{LOG10}(k) = ((\mathrm{a} - \mathrm{b} \times \mathrm{LOG10(RFN)}) + (\mathrm{c} - \mathrm{d} \times \mathrm{LOG10(RFN)}) \times \mathrm{LOG10(ippor)}), \quad (1)$$

where RFN = rock-fabric number; ippor = interparticle porosity; $a = 9.7982$; $b = 12.0838$; $c = 8.6711$; and $d = 8.2965$.

Normally, interparticle porosity is used along with rock-fabric number (RFN) in equation 1 to calculate permeability. However, the limited log suites available throughout the field do not allow for calculation of interparticle porosity. Therefore, as discussed by Harrington and Lucia (2012), total porosity (including interparticle and vug porosity) was used in this study. When total porosity is used, the term "apparent rock-fabric number" (ARFN) is used to indicate that an assumption is made equating total porosity to interparticle porosity. In this study, ARFN was used in the global transform along with total porosity (equation 2). The ARFN is provided by the relationship between ARFN and stratigraphy, as discussed by Harrington and Lucia (2011), and total porosity is from porosity logs (Kane and Jennings, 2011).

$$\mathrm{LOG10}(k) = ((\mathrm{a} - \mathrm{b} \times \mathrm{LOG10(ARFN)}) + (\mathrm{c} - \mathrm{d} \times \mathrm{LOG10(ARFN)}) \times \mathrm{LOG10(phi)}), \quad (2)$$

where ARFN = apparent rock-fabric number; phi = total porosity; $a = 9.7982$; $b = 12.0838$; $c = 8.6711$; and $d = 8.2965$.

The method was tested on six cored wells in the simulation area (Figure 1) with good to excellent results. Lower Clear Fork rocks assigned to sequence L2.2 (sequence designations defined by Ruppel and Harrington, 2012) are composed of class 2 anhydritic dolostone except for a small area of the field where L2.2 is limestone. A porosity-permeability crossplot of data from the anhydritic dolostone (Figure 2A) shows that most data fall within the class 1 field. As discussed by Harrington and Lucia (2012), class 2 dolostones plot in the class 1 field because of the presence

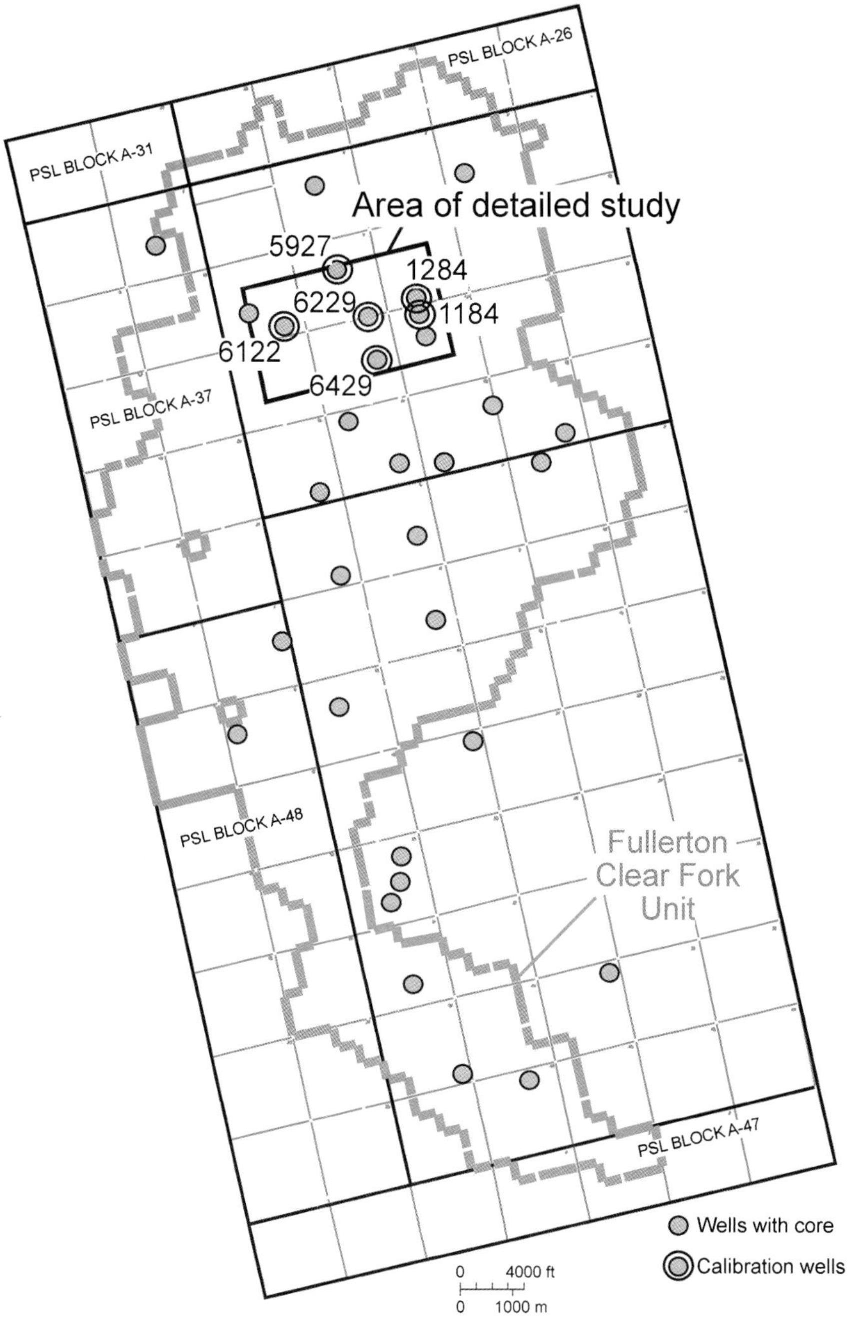

FIGURE 1. Map showing area of detailed study and location of calibration wells.

of large amounts of poikilotopic anhydrite. The match between core and calculated permeability values is excellent using an ARFN of 1 (Figure 2B).

As shown by Harrington and Lucia (2012), sequence L2.2 is dominantly limestone in a small area in the northern part of the study area. Porosity and permeability data from these rocks plot along the class 2-3 boundary, indicating an ARFN of 2.5 (Figure 3A). The data plot along the class 2-3 boundary using total porosity because the porous limestone is dominated by moldic class 1 grainstone and class 2 grain-dominated packstones and grain molds do not add significantly to permeability. The grainstones would plot in the class 1 field if moldic porosity was subtracted from total porosity. As shown in Figure 3B, the match between core and calculated permeability values is quite good using an ARNF of 2.5 except for the upper 10 ft (3 m) where the high permeability is the result of a different fabric, a class 1 large-crystalline dolostone.

Lower Clear Fork rocks in the underlying sequence L2.1 comprise both peritidal and subtidal facies (Figure 4). The upper part of sequence L2.1 is composed of class 3 fine-crystalline, mud-dominated peritidal dolostones that are characterized by an ARFN

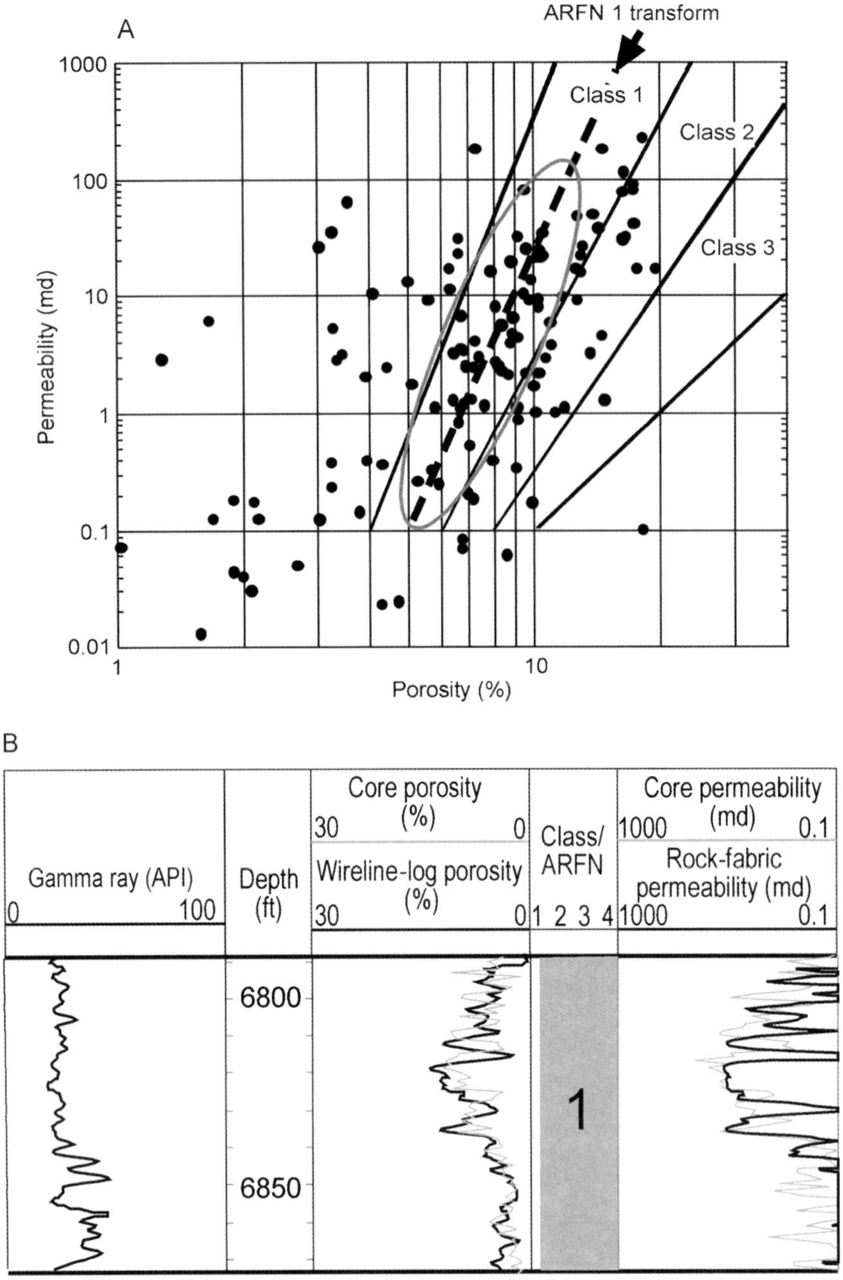

Figure 2. Permeability estimation for dolostone in sequence L2.2. (A) Porosity-permeability crossplot showing that most data points fall in the class 1 field. Accordingly, a class 1 transform was used in modeling. (B) Comparison of calculated permeability with core permeability in the dolostone facies of sequence L2.2 using a class 1 transform. Note that the calculated values closely fit core permeability data except in low-porosity intervals. Well FCU 6229. ARFN = apparent rock-fabric number; FCU = Fullerton Clear Fork Unit.

of 3 (Figure 4). The lower part of the sequence is composed of subtidal dolostones and limestones that cannot be readily subdivided stratigraphically or geographically. The porous limestones are moldic grain-dominated fabrics that tend to plot in the class 2 field. The mud-dominated limestones are very low in permeability. In general, the dolostones are class 2 medium-crystalline, grain-dominated dolopackstones and wackestones (see Harrington and Lucia, 2012). As a result, an ARFN of 2 was used to calculate the permeability of these rocks. As illustrated in Figure 4, a good match between core and calculated permeability is observed.

The Wichita interval (sequences L2.0 and L1.2) is dominated by porous fine-crystalline dolostones and tight mud-dominated limestone of the peritidal facies characterized by an ARFN of 3. A good correlation exists between core permeability and calculated permeability in sequence L2, except for a few permeability spikes that are assumed to be sample bias (Figure 5A). In sequence L1.2, log calculations tend to underestimate permeability. Core descriptions and one image log show collapse breccias that developed locally in this interval (Figure 5B), suggesting the presence of fracturing caused by karsting and collapse that may account for higher than predicted permeability in this

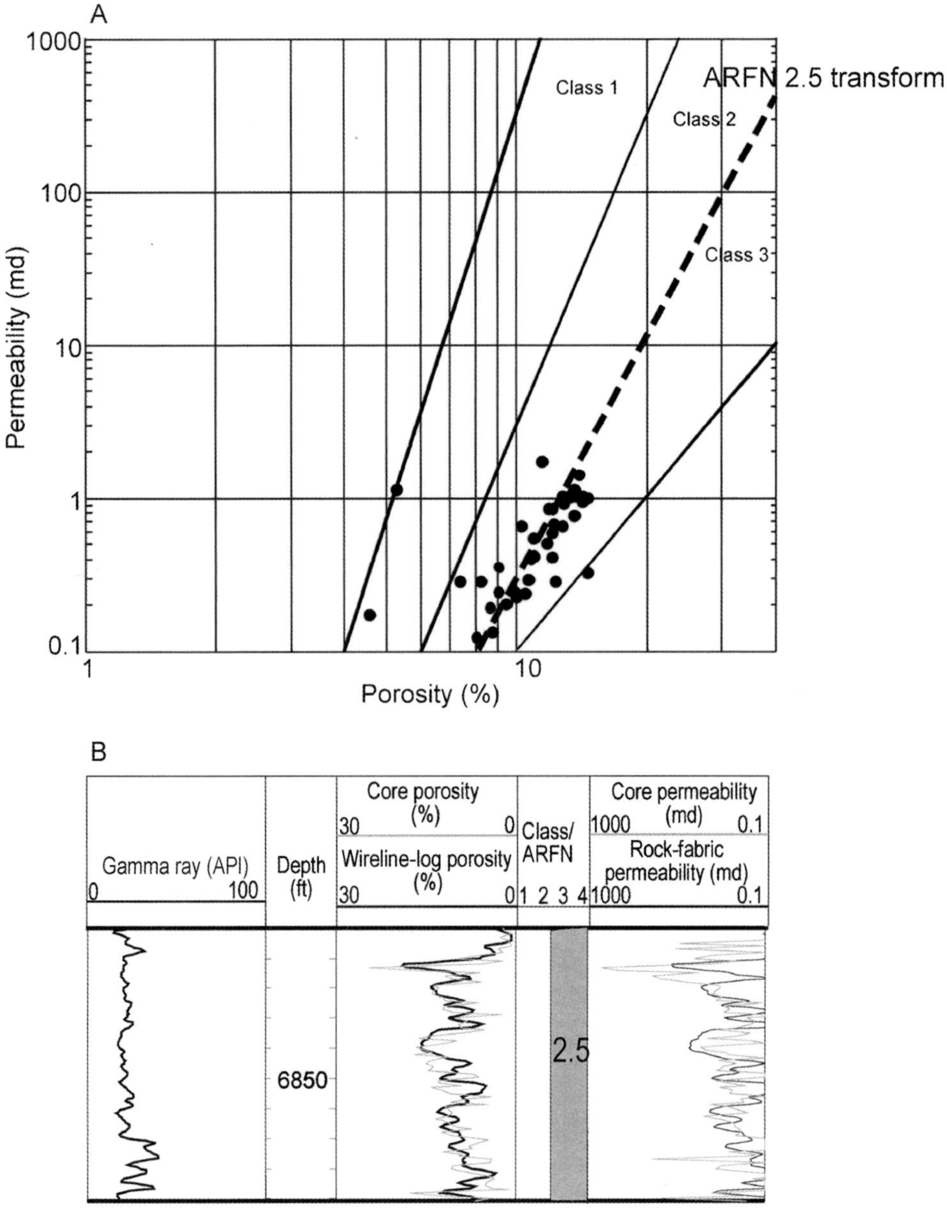

FIGURE 3. Permeability estimation for limestone in sequence L2.2. (A) Porosity-permeability crossplot for moldic limestone in sequence L2.2. The data cluster along the class 2-3 boundary because moldic porosity adds little to permeability. (B) Comparison of calculated permeability with core permeability in the limestone facies of sequence L2.2 using an apparent rock-fabric number (ARFN) 2.5 transform. The high permeability in the upper 10 ft (3 m) is caused by the local presence of class 1 large-crystalline dolostone. Well FCU 6122. FCU = Fullerton Clear Fork Unit.

area. As discussed by Wang and Lucia (2012), matrix permeability estimates were increased about 20 times to match historical data on injection volumes in the simulation model.

The crossplots of core permeability and log-calculated permeability of these four data sets are illustrated in Figure 6, along with power-law statistical equations. The correlation coefficient (*R*) for the anhydritic dolostone in sequence L2.2 (Figure 6A) is the highest of the group at 0.75, whereas the other three plots have *R* values of about 0.3. All plots show considerable scatter. However, by inspection, most of the data points that fall within a factor of 10 envelop around the statistical relationship. It has been shown (Lucia, 1999; Jennings et al., 2000) that permeability in carbonates commonly ranges by a factor of 10 or more on a scale of inches or feet with little spatial organization. Also, these data are from logs and core plugs, which are at very different scales. Wireline-log data and calculated permeabilities are averaged over about 2 ft (~0.6 m), whereas the core data are from 1 in. (2.54 cm) plugs. In addition, core depths and log depths commonly differ considerably. As a result, core depths must be shifted to log depths for comparison, and that shift is always subject to inaccuracies. Therefore, permeability values that are being compared may not be from the same depth. Core and calculated permeabilities appear to be better correlated in depth plots (Figures 2–5) because scale and depth problems are less apparent to the observer. Therefore, a scatter of this magnitude is

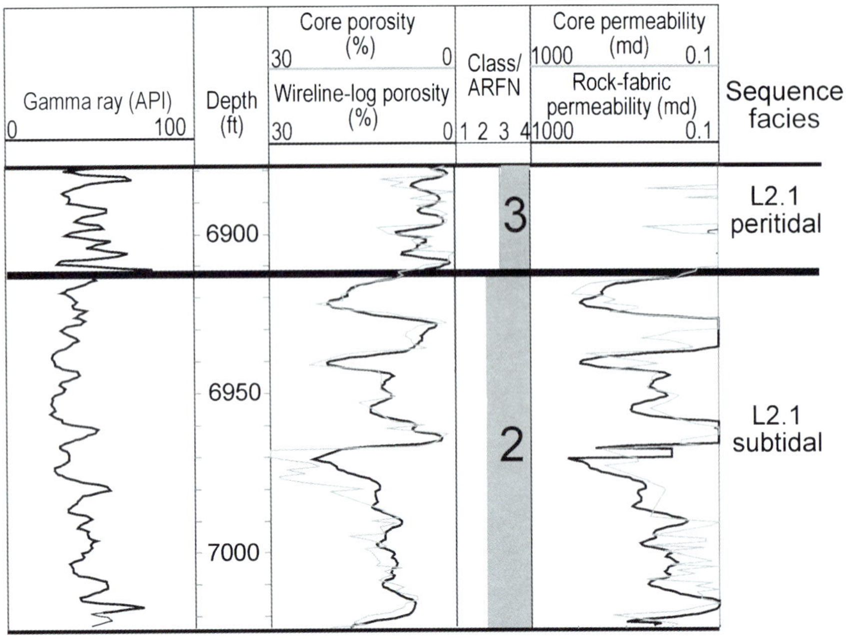

FIGURE 4. Comparison of calculated permeability with core permeability in sequence L2.1. A class 3 transform is used in the upper peritidal interval, and a class 2 transform, in the lower subtidal interval. Well FCU 6429. ARFN = apparent rock-fabric number; FCU = Fullerton Clear Fork Unit.

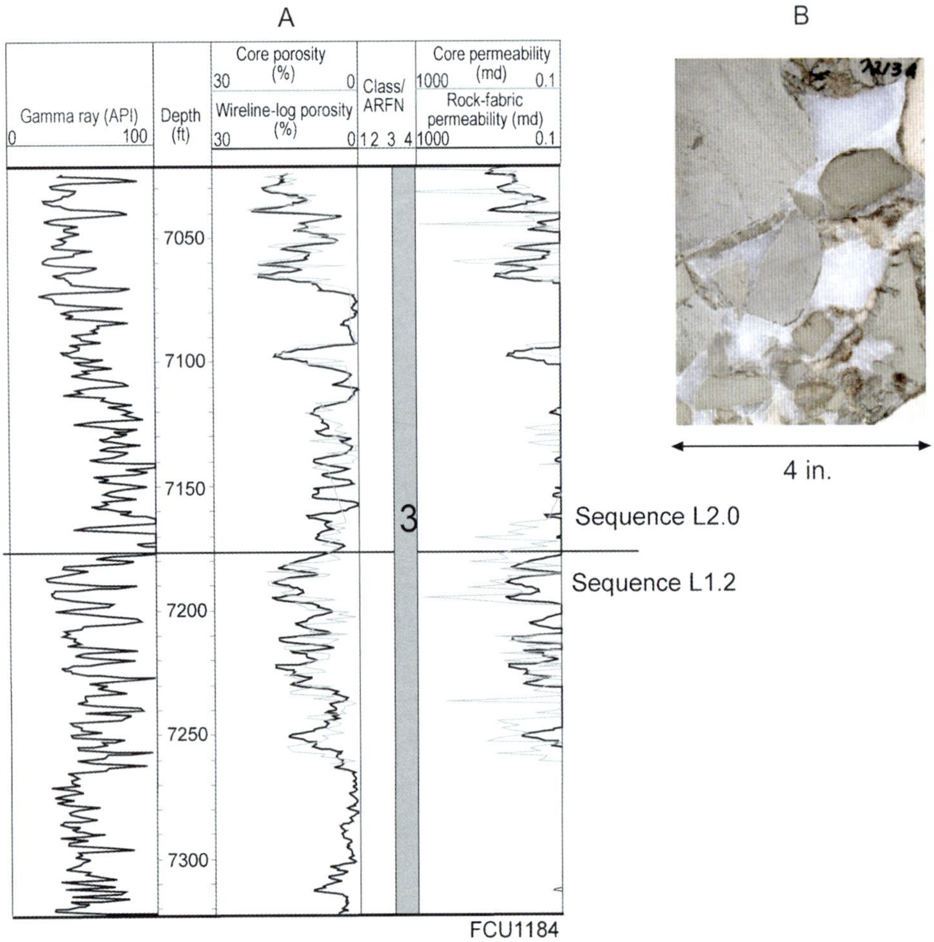

FIGURE 5. Permeability in Wichita peritidal facies. (A) Comparison of calculated permeability with core permeability showing good correlation in sequence L2 and poor correlation in L1. (B) A core photograph of karst breccia found in L1 suggesting the presence of a touching-vug pore system.

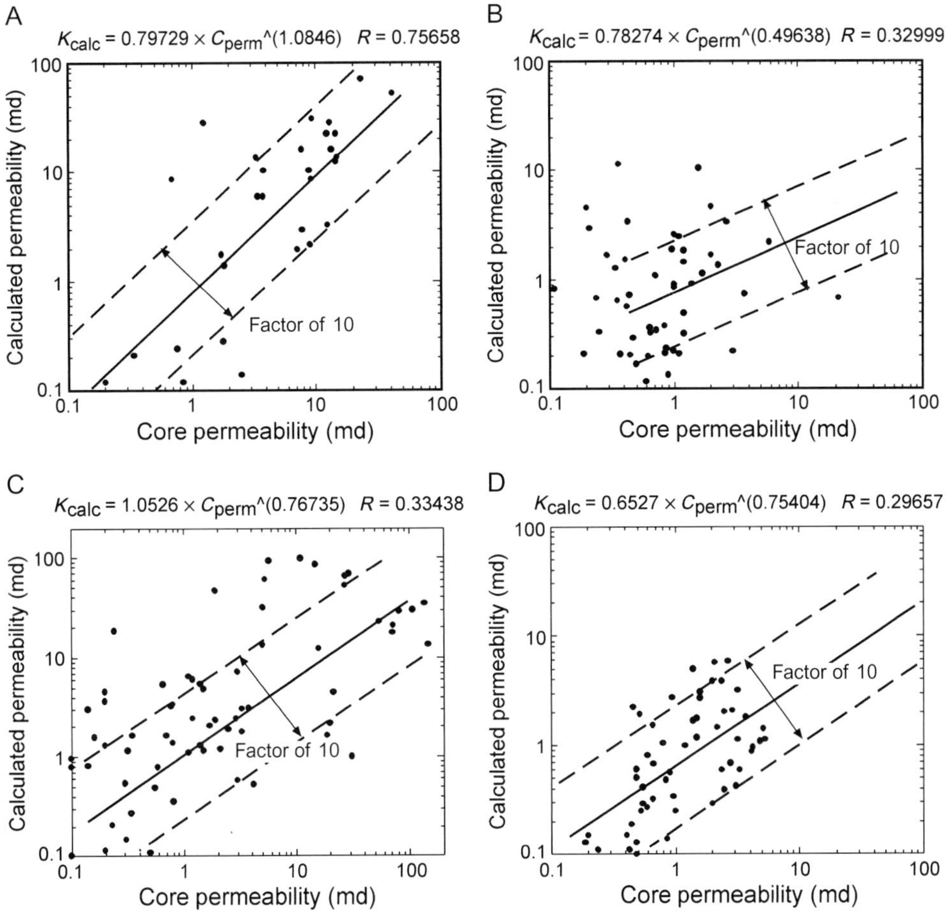

FIGURE 6. Comparison of calculated permeability and core permeability values. (A) Sequence L2.2 dolostone in well FCU 6229. (B) Sequence L2.2 limestone in well FCU 6122. (C) Sequence L2.1 dolostone and limestone in well FCU 6229. (D) Wichita dolostone in sequence L2 of well FCU 1184. ARFN = apparent rock-fabric number; FCU = Fullerton Clear Fork Unit.

to be expected, and the calculated permeabilities are considered adequate for characterizing this reservoir.

ESTIMATION OF ORIGINAL WATER SATURATION

Normally, initial water saturation is calculated from porosity and resistivity logs. In the Fullerton Clear Fork field, however, no resistivity logs are available for most of the wells. Therefore, estimates of initial water saturation (S_{wi}) were made using relationships among saturation, reservoir height, and petrophysical class based on capillary-pressure data. The conventional method for estimating S_{wi} using capillary-pressure curves is the Leverett J function, which relates saturation to reservoir height and the ratio of permeability and porosity (*k*/phi). The difficulty with this method is that a permeability value is required. Permeability is commonly obtained from a single porosity-permeability transform, which results in S_{wi} being a function of porosity and reservoir height. However, it is well known that S_{wi} is a function of pore size as well as porosity and reservoir height.

Lucia (1995) showed that the combination of rock fabric and interparticle porosity is a measure of pore size. By dividing capillary-pressure curves into petrophysical classes and grouping the curves into porosity bins, we can show that S_{wi} is a function of petrophysical class, porosity, and reservoir height. Resulting equations are given as follows:

$$\text{class 1 } S_{wi} = 0.02219 \times H^{-0.316} \times \phi^{-1.745}, \quad (3)$$

$$\text{class 2 } S_{wi} = 0.1404 \times H^{-0.407} \times \phi^{-1.440}, \quad (4)$$

$$\text{class 3 } S_{wi} = 0.6110 \times H^{-0.505} \times \phi^{-1.210}, \quad (5)$$

where S_{wi} = initial water saturation in fraction; H = reservoir height in feet; and ϕ = porosity in fraction (assuming that little vuggy porosity is present).

The difficulty with this method is the requirement that the petrophysical class be determined. In this study, petrophysical class was provided from ARFN obtained from rock-fabric stratigraphy. Apparent rock-fabric numbers 1, 2, or 3 are equivalent to petrophysical classes 1, 2, and 3, respectively. The porosity is

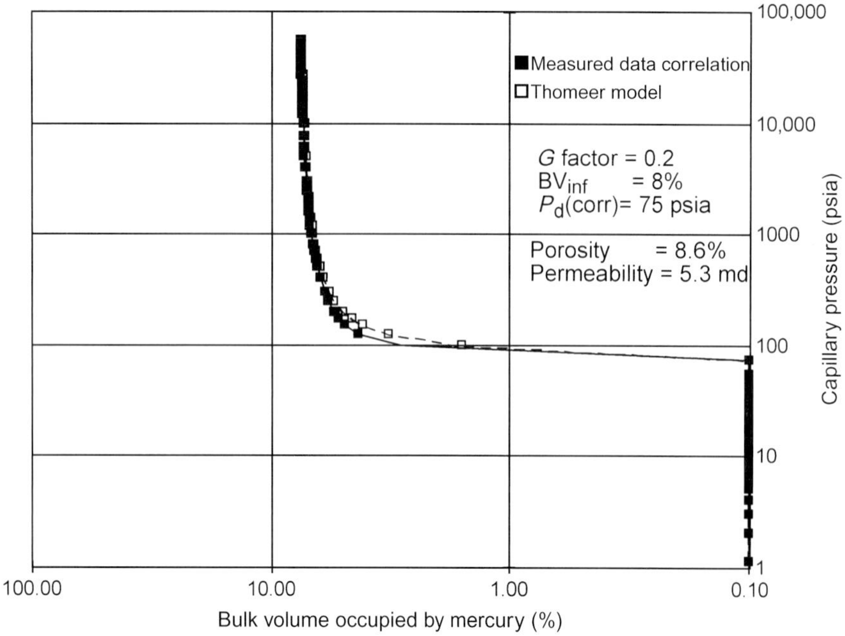

FIGURE 7. An example of fitting Thomeer parameters to a class 2 medium-crystalline dolostone. BV_{inf} = fraction of bulk volume occupied by mercury extrapolated to infinite pressure; G factor = Thomeer parameter related to the shape of the capillary-pressure curve; P_d = entry capillary pressure corrected for surface effects in psi.

from wireline logs, and the reservoir height is calculated based on a zero capillary-pressure level below the Wichita at −4100 ft (−1250 m). The Wichita and Lower Clear Fork are assumed to be one reservoir. Multiple oil-water contacts have been suggested in past field studies, but evidence presented on multiple oil-water contacts is thought to be equivocal. A good evidence of an oil-water contact is observed at about −4100 ft (−1250 m) from log and core data in a deep well on the margin of the field.

During the course of the Fullerton study, new capillary-pressure data were obtained from samples selected on the basis of rock-fabric descriptions and porosity. The new capillary-pressure data indicate that class 2 and 3 generic saturation models overestimate water saturation, suggesting the need for revised models.

We developed new models using the method of Thomeer (1960) of characterizing capillary-pressure curves. The Thomeer (1960) method estimates irreducible water saturation, entry pressure, and shape of the capillary-pressure curve (the G factor) from a crossplot of percentage bulk volume of mercury occupied and entry pressure corrected for surface effects:

$$(BV_{occ}/BV_{inf}) = e\hat{}(-G/(LOG(P_c/P_d))), \qquad (6)$$

where BV_{occ} = fraction of bulk volume occupied by mercury in fraction; BV_{inf} = fraction of bulk volume occupied by mercury extrapolated to infinite pressure; G = Thomeer parameter related to the shape of the capillary-pressure curve; P_c = capillary pressure in psi; and P_d = entry capillary pressure corrected for surface effects in psi.

The values for the above parameters were selected for each capillary-pressure curve until a reasonable match was made between the data and the model using a spreadsheet program (Figure 7). In several cases, a bimodal fit was required; the fit to the lowest pressure data was used in the new model. Injection pressures of greater than 50,000 psi (>345 MPa) were used in the measurements, resulting in the BV_{inf} value equaling total porosity. Therefore, BV_{occ}/BV_{inf} is assumed to be S_{wi}. Samples were then grouped by petrophysical class, and porosity was correlated to the G factor and to the extrapolated entry pressure. Data are presented in Tables 1 and 2. The correlation between porosity and G factor for classes 2 and 3 is poor (Figure 8), and the difference in G factor values is small. The average value of class 2 is 0.22, and the average value of class 3 is 0.14 (excluding two anomalously high B factors). Extrapolated entry pressure correlated well with porosity for both class 2 and 3 rocks, although correlation was different in each case (Figure 9).

$$\text{class 2 } P_d = 0.0768067 \times \text{Phi}\hat{}(-2.8394)\ R = 0.7 \qquad (7)$$

$$\text{class 3} P_d = 3.1117 \times \text{Phi}\hat{}(-1.9717)\ R = 0.8 \qquad (8)$$

The resulting new saturation models for classes 2 and 3 are given below. Equation 6 is the basic Thomeer equation, which can be written in terms of S_{wi} as:

$$S_{wi} = 1 - [e\hat{}(-G/(LOG10(P_c/P_d)))], \qquad (9)$$

Table 1. Thomeer and core-analysis values for class 2 samples.

Depth	*Sample Number*	*Well FCU**	*G Factor*	P_d* *(psi)*	BV_{inf}*	*Phi* (%)*	*Permeability (md)*	*Rock Fabric*	*Petrophysical Class*
6822	10	6429	0.10	150	1.5	10.7	1.13	MxlMDDstn*	2
6855	42	6429	0.10	100	2.0	14.4	5.60	GDDP*	2
6907	94	6429	0.20	75	8.0	8.6	0.53	MxlMDDstn*	2
6908	95	6429	0.30	60	10.0	10.1	8.30	MxlMDDstn*	2
6991	98	5927	0.30	60	8.0	8.4	0.28	MxlMDDstn*	2
6995	102	5927	0.20	25	12.0	12.8	9.76	MxlMDDstn*	2
6813	1	6429	0.30	10	18.0	12.5	9.63	GDDP*	2
6818	6	6429	0.25	80	9.0	11.6	2.38	GDDP*	2
6914	101	6429	0.20	20	18.0	17.5	16.23	GDDP*	2
6915	102a	6429	0.30	40	18.0	18.5	16.30	GDDP*	2
6938	57	5927	0.20	100	8.0	8.0	0.58	MxlMDDstn*	2

**FCU = Fullerton Clear Fork Unit; P_d = entry capillary pressure corrected for surface effects in psi; BV_{inf} = fraction of bulk volume occupied by mercury extrapolated to infinite pressure; Phi = total porosity; MxlMDDstn = medium crystalline mud-dominated dolostone; GDDP = grain-dominated dolopackstone; G Factor = thomeer pore geometrical factor.*

which assumes that total porosity is filled with mercury at infinite pressure.

Mercury capillary pressure is converted to reservoir height using:

$$\frac{H}{P_c} = \frac{(\sigma \cos\theta)_{o/w/s}}{0.434(\rho w - \rho o) \times (\sigma \cos\theta)_{hg/a/s}} \tag{10}$$

and using the data from the Fullerton Clear Fork reservoir as presented in Table 3; $P_c = 1.19$ H.

The class 2 saturation model (equations 11, 12) was developed using a *G* factor of 0.2 and equation 7:

$$S_{wi}\ (\text{class } 2) = 1 - [e\text{^}(-0.2/(\text{LOG10}(P_c /(0.0768067 \times \text{Phi}\text{^}(-2.8394)))))]. \tag{11}$$

In terms of reservoir height (*H*):

$$S_{wi}(\text{class } 2) = 1 - [e\text{^}(-0.2/(\text{LOG10}(15.5064 \times H/\text{Phi}\text{^}(-2.8394))))]. \tag{12}$$

Table 2. Thomeer and core-analysis values for class 3 samples.

Depth	*Sample Number*	*Well FCU**	*G Factor*	P_d* *(psi)*	BV_{inf}*	*Phi* (%)*	*Permeability (md)*	*Rock Fabric*	*Petrophysical Class*
7001	108	5927	0.12	150	17	17.9	0.990	FxlMDDstn*	3
7002	109	5927	0.15	56	23	23.6	5.660	FxlMDDstn*	3
7003	110	5927	0.12	140	14	14.2	0.600	FxlMDDstn*	3
7004	111	5927	0.13	300	3	20.1	0.880	FxlMDDstn*	3
7007	114	5927	0.13	150	5	26.2	15.90	FxlMDDstn*	3
7009	116	5927	0.15	120	15	15.7	0.950	FxlMDDstn*	3
7044	147	5927	0.13	80	18	18.6	3.790	FxlMDDstn*	3
7045	148	5927	0.13	100	19	19.1	2.640	FxlMDDstn*	3
7054	157	5927	0.20	175	10	10.5	0.232	FxlMDDstn*	3
7056	159	5927	0.13	100	19	19.5	1.930	FxlMDDstn*	3
7059	162	5927	0.13	100	15	15.9	0.450	FxlMDDstn*	3
7089	192	5927	0.15	200	13	13.3	0.267	FxlMDDstn*	3

**FCU = Fullerton Clear Fork Unit; P_d = entry capillary pressure corrected for surface effects in psi; BV_{inf} = fraction of bulk volume occupied by mercury extrapolated to infinite pressure; Phi = total porosity; FxlMDDstn = fine-crystalline mud-dominated dolostone.*

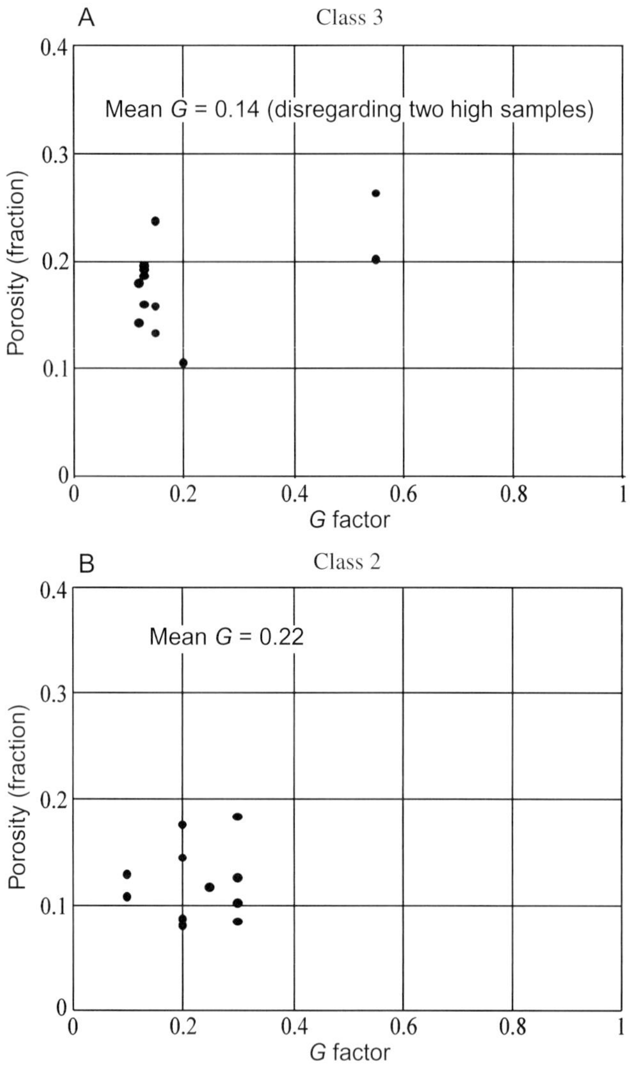

FIGURE 8. Crossplots of Thomeer *G* factor and porosity for (A) class 3 fabrics and (B) class 2 fabrics showing little relationship between porosity and *G* factor. The two class 3 samples with high *G* factors are bimodal and require two curves to fit. *G* factor = Thomeer parameter related to the shape of the capillary-pressure curve.

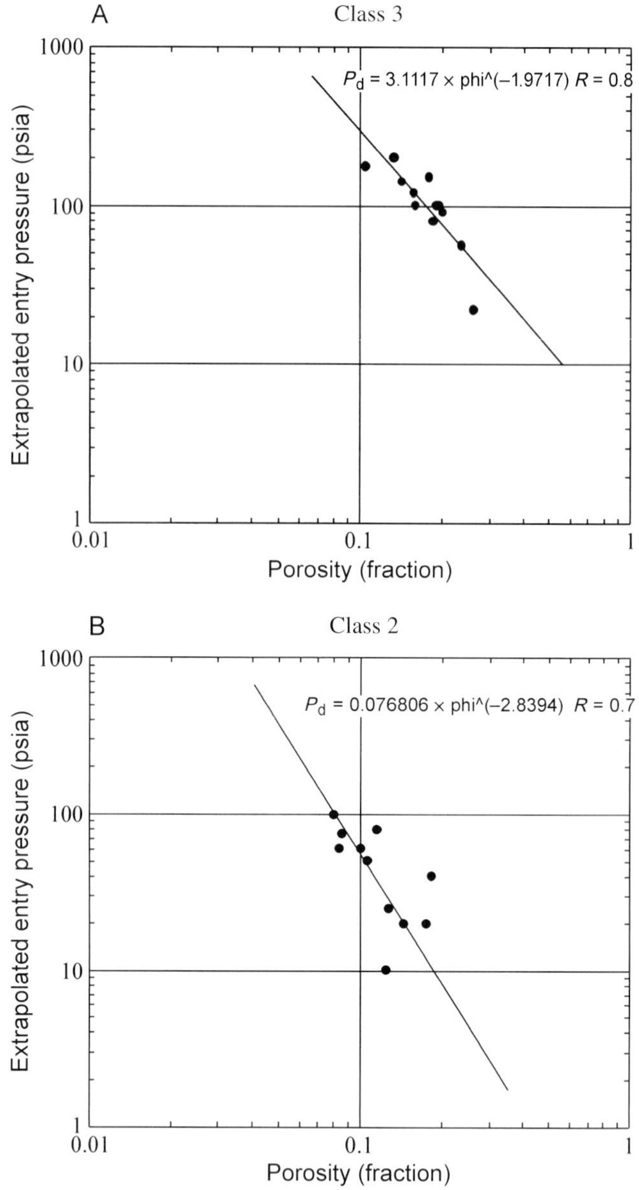

FIGURE 9. Crossplots of extrapolated entry pressures and porosity for (A) class 3 fabrics and (B) class 2 fabrics showing statistical relationships between entry pressure and porosity. P_d = entry capillary pressure corrected for surface effects in psi; phi = total porosity.

The class 3 saturation model (equations 13, 14) was developed using a *G* factor of 0.1 and equation 8.

$$S_{wi}(\text{class 3}) = 1 - [\text{e}\hat{}(-0.1/(\text{LOG10}(P_c /(3.1117 \times \text{Phi}\hat{}(-1.9717)))))]. \quad (13)$$

In terms of reservoir height (*H*):

$$S_{wi}(\text{class 3}) = 1 - [\text{e}\hat{}(-0.1/(\text{LOG10}(0.3827 \times H/(\text{Phi}\hat{}(-1.9717)))))]. \quad (14)$$

The new capillary-pressure models were compared with capillary-pressure data and with the original

Table 3. List of parameters used to convert laboratory mercury capillary-pressure measurements to reservoir conditions.

	Reservoir: Oil/Water	*Laboratory: Mercury/Air*
Interfacial tension	30	480
Contact angle	30	140
Oil density	0.820	
Water density	1.015	

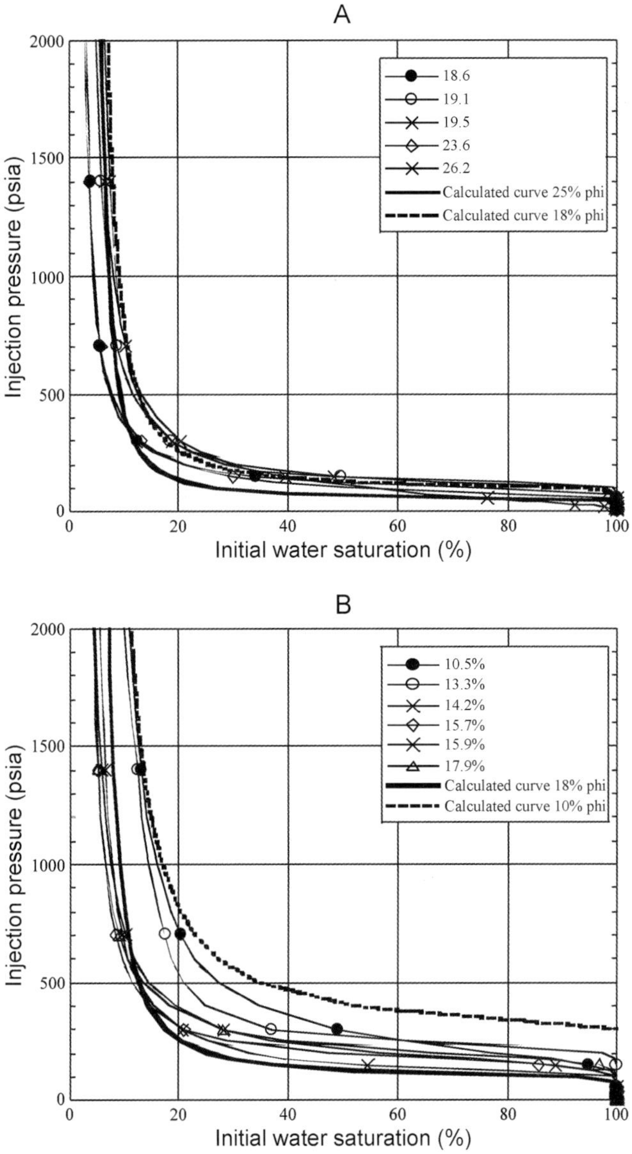

FIGURE 10. **Comparison of class 3 curves with curves calculated using the Thomeer model. (A) Porosity less than 18% and (B) porosity greater than 18%. Phi = total porosity. Phi = porosity.**

generic models by visual inspection. The new class 3 model compares well with the data (Figure 10). Above 18% porosity, the model results cluster with the data (Figure 10A), and below 18% porosity, the model results bracket the data (Figure 10B). Below 40% S_{wi}, the new model significantly calculates a lower S_{wi} value than the generic model (Figure 11A, B).

The new class 2 model also compares well with the data (Figure 12). The model matches two of the three curves with 8% porosity (Figure 12A), straddles the data with 10 to 14% porosity (Figure 12B), and matches one of the two curves with about 18% porosity (Figure 12C). It significantly calculates lower S_{wi} values than the generic model below 20% S_{wi} (Figure 13A, B).

Six capillary-pressure curves from the anhydritic dolostone in sequence L2.2 were obtained and are illustrated in Figure 14. Although this is a class 2 medium-crystalline dolostone, a class 1 capillary-pressure model was used because the data plot in the class 1 field (Figure 2). A good comparison between the generic class 1 model and the data exists (Figure 14). Therefore, a generic class 1 model was used to calculate S_{wi} for the anhydritic dolostone in L2.2, and no new model was needed.

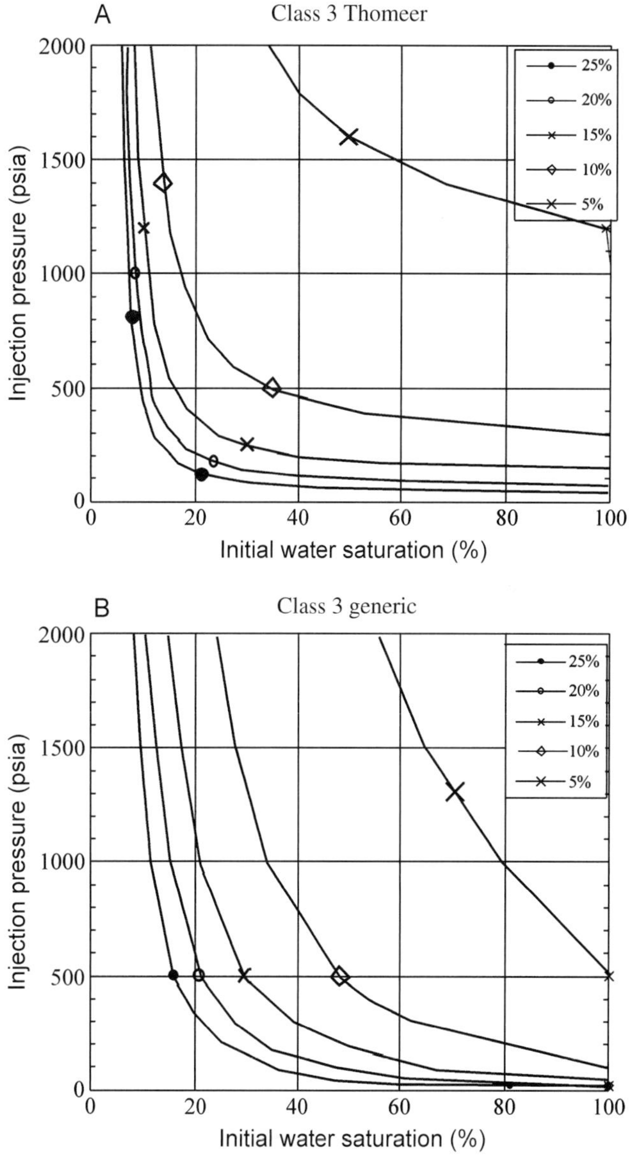

FIGURE 11. **Comparison of (A) Thomeer class 3 capillary-pressure model with (B) generic class 3 model. The Thomeer model predicts lower water saturation below 40% saturation.**

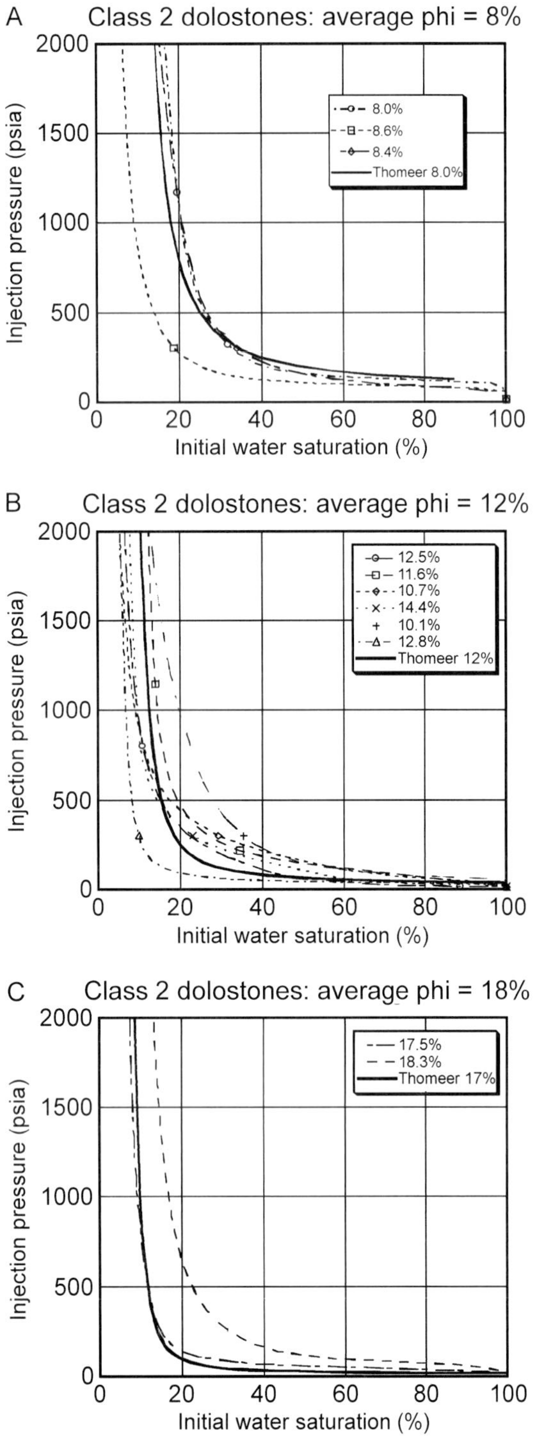

FIGURE 12. Comparison of class 2 curves with curves calculated using the Thomeer model. (A) Porosity average = 8%; (B) porosity average = 12%; (C) porosity average = 18%. Phi = total porosity.

Capillary-pressure curves for moldic grainstones are typically difficult to model because of their bimodal pore structure. Interestingly, the oomoldic grainstone capillary-pressure curves from the Fullerton Clear Fork field, illustrated in Figure 15, are not bimodal. Visual comparison suggests that they are most similar to class 3 curves calculated from the new model (Figure 11A). The class 3 model was selected to estimate S_{wi} in the relatively small area where sequence L2.2 is limestone. In sequence L2.1, however, oomoldic grainstones are interbedded with class 2 dolostones and were not mapped separately. A decision was made to use the class 2 model to estimate S_{wi} because dolostone is the predominant fabric. Using a class 2 model instead of a class 3 model results in underestimating S_{wi} for oomoldic grainstones interbedded with class 2 dolostones. This is because a class 3 model calculates higher water saturation values than a class 2 model, given equal porosity values.

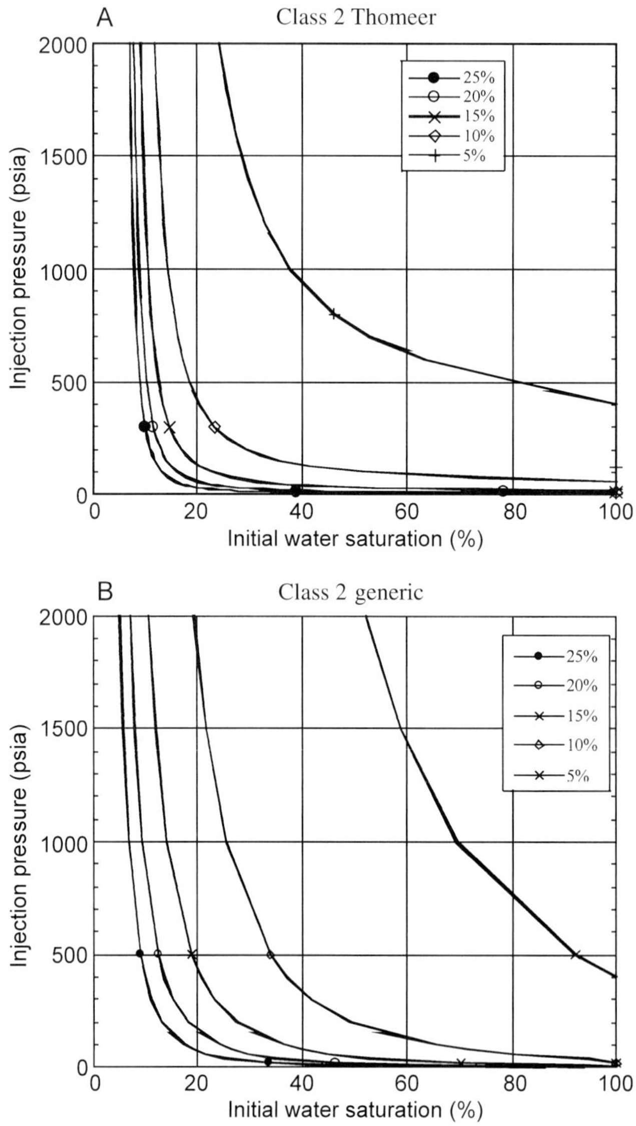

FIGURE 13. Comparison of (A) Thomeer class 2 capillary-pressure model with (B) generic class 2 model. The Thomeer model predicts lower water saturation below 20% saturation.

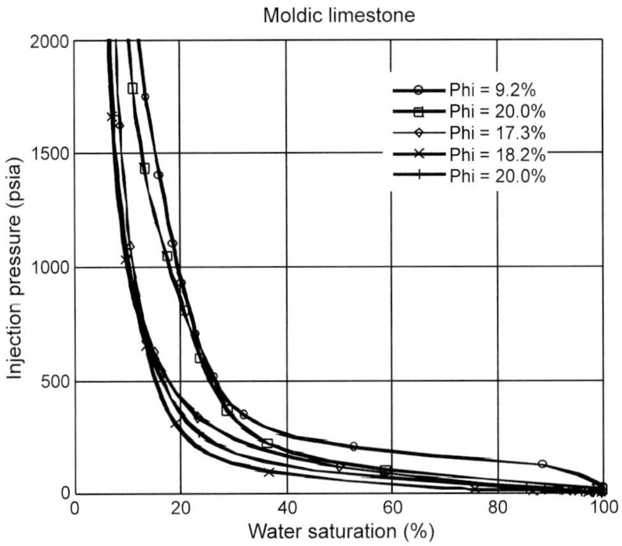

FIGURE 14. Comparison of capillary-pressure curves from class 2 anhydritic dolostones that plot in the class 1 field with curves using the generic class 1 model. Phi = porosity.

The new class 2 and 3 capillary-pressure curves match the data better than the generic curves and result in lower S_{wi} values. The class 2 anhydritic dolostone that plots in the class 1 field is best modeled using the generic class 1 saturation model. The S_{wi} was calculated for each well using these saturation models, imputing porosity from wireline logs and petrophysical class from stratigraphy.

COMPARISON OF MODEL INITIAL WATER SATURATION WITH ARCHIE WATER SATURATION

To test the accuracy of our capillary-pressure-based saturation models, calculated water saturation values were compared with values calculated using the Archie equation and resistivity logs. Archie water saturation values were calculated using m and n of 2, which are supported by laboratory measurements given in Table 4. Water resistivity was estimated at 0.04 ohm m.

Waterflooding was initiated in 1961, and most of the field was under waterflood by 1973. As a result, capillary-pressure S_{wi} values are expected to be equal to or lower than Archie values calculated from logs acquired after 1973. The well illustrated in Figure 16 was drilled after 1973 and shows intervals of both class 2 and class 3 fabrics, in which the model S_{wi} and Archie water saturation match closely, suggesting that the capillary-pressure models give reasonable estimates of S_{wi}. However, in some intervals, for example, at 7010 ft (2137 m) (Figure 16), Archie water saturation is higher than model S_{wi}. This well was drilled after the initiation of waterflooding, and these intervals are interpreted to contain injection water. Because injection waters are commonly less saline and more resistive than 0.04 ohm m, Archie S_{wi} is expected to underestimate S_{wi} in flooded intervals. Archie water saturation is also much higher than model S_{wi} in the upper 20 ft (6 m) of the interval shown in Figure 16. However, this is caused by the very low porosity in this part of the well. The Archie S_{wi} values are typically unreliable where porosity is less than 5 or 6%.

DISCUSSION

The stratigraphic approach used to predict ARFN results in reasonable estimates of matrix permeability using the modified global rock-fabric transform equation. The incorporation of fundamental variations in rock fabrics throughout the reservoir and the relationship between porosity and permeability within units with constant ARFNs provide a firm basis for modeling the permeability structure of this reservoir. Good results are achieved despite the complex interbedding of limestones and dolostones and grain- and mud-dominated fabrics. The presence of grain molds in the grainstone causes data points for these rocks to plot into the class 2 field, and the conversion of limestone into a medium-crystalline dolostone causes mud-dominated dolostones to plot in the class 2 field. The result of these diagenetic complexities is that these mixed rock fabrics tend to all plot in the class 2 field

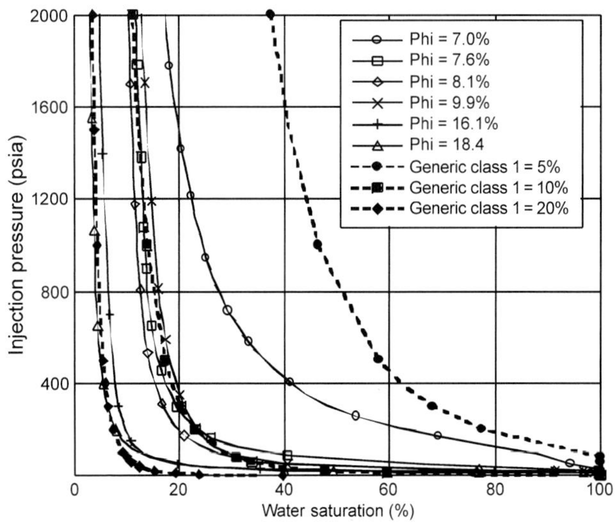

FIGURE 15. Capillary pressure curves from moldic grainstones. Phi = porosity.

Table 4. List of electrical measurements from the Lower Clear Fork.

Well	*Formation*	*Sample Number*	*Rock Fabric*	*Separate Vug (%)*	*Petrophysical Class*	*m*	*n*
FCU* 6429	LCF*	1	GDDP*	2	2	2.05	2.04
FCU* 6429	LCF*	6	GDDP*	3	2	2.22	2.20
FCU* 6429	LCF*	8	GDDP*	2	1	1.86	1.97
FCU* 6429	LCF*	20	GDDP*	0	1	1.86	1.94
FCU* 5927	LCF*	28	Moldic grainstone	12	1	2.04	2.10
FCU* 5927	LCF*	31	Moldic grainstone	20	1	2.14	2.11
FCU* 5927	LCF*	32	Moldic grainstone	15	1	2.13	2.02
FCU* 6429	LCF*	32a	GDDP*	0	1	1.85	1.91
FCU* 5927	LCF*	35	Moldic grainstone	7	1	2.05	1.95
FCU* 6429	LCF*	35a	GDDP*	2	1	1.89	1.85
FCU* 5927	LCF*	57	MDDP*	1	2	1.83	1.87
FCU* 6429	LCF*	100	GDDP*	0	1	1.84	1.91
FCU* 6429	LCF*	101	GDDP*	1	2	1.89	1.94
FCU* 6429	LCF*	102a	GDDP*	0	2	1.88	1.89
					Average	1.96	1.98

**FCU = Fullerton Clear Fork Unit; LCF = Lower Clear Fork; GDDP = grain-dominated dolopackstone; MDDP = mud-dominated dolopackstone; m = Archie cementation exponent; N = Archie saturation exponent.*

and can thus be characterized by an ARFN of 2. Peritidal facies are typically class 3 fine-crystalline dolostones and can be characterized by an ARFN of 3. Mud-dominated limestones associated with peritidal dolostone are typically low in permeability and calculate low permeability whether an ARFN of 2 or 3 is used. In sequence L1, however, there appears to be a touching-vug pore system, and permeability values based on the global transform are understated.

The new capillary-pressure models developed for calculating S_{wi} in class 2 and 3 fabrics fit the capillary-pressure data better than generic rock-fabric models. The new models are based on the Thomeer method, which relates S_{wi} to an extrapolated entry pressure, a curve-shaped factor (G), and total porosity. Like the generic models, S_{wi} is related to petrophysical class, porosity, and reservoir height. Analysis of input parameters indicates that extrapolated entry pressure

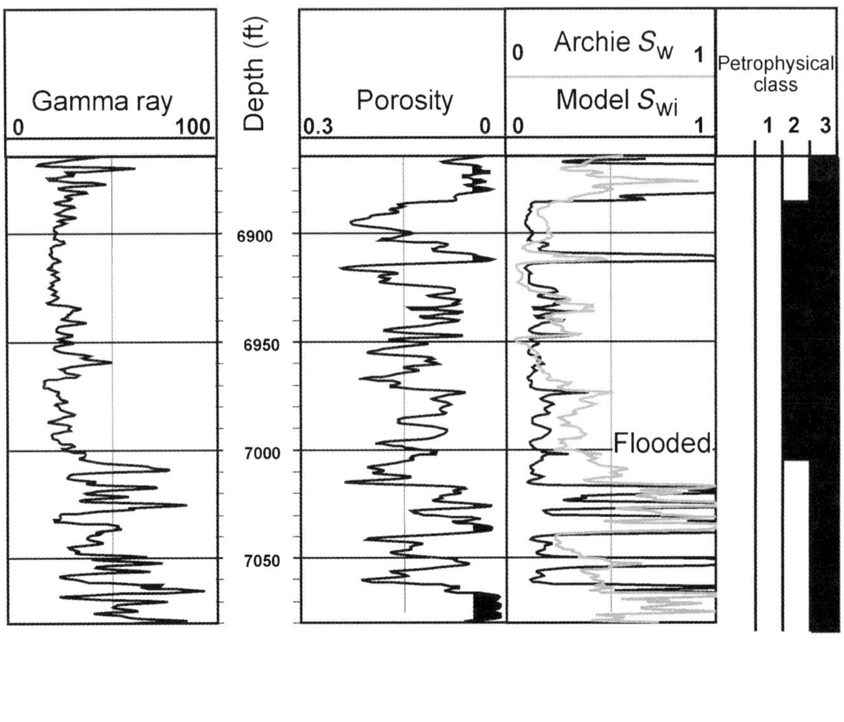

FIGURE 16. Example of initial water saturation (S_{wi}) from capillary-pressure models compared with Archie water saturation (S_w). The areas where Archie S_w is much higher than S_{wi} are thought to be flooded intervals. Well FCU 5727. FCU = Fullerton Clear Fork Unit.

varies with porosity and petrophysical class and that the G factor is poorly correlated with porosity and is slightly different between classes 2 and 3. The zero capillary-pressure level of −4100 ft (−1250 m) is based on limited data and assumes that the Wichita and Lower Clear Fork are a single reservoir. Comparisons with Archie water saturation calculations from wireline resistivity logs suggest that the model saturations are reasonable.

The results of this study illustrate the value of integrating stratigraphy and diagenesis with petrophysical measurements for calculating permeability and initial saturation from wireline logs for use in the construction of static reservoir models. Petrophysical parameters are a function of porosity and pore size. In this study, porosity is obtained from wireline logs, and pore size is obtained from rock fabrics. In carbonate fields where adequate wireline logs are available, rock-fabric numbers can be obtained from log calculations. However, in many old fields, only gamma-ray and neutron logs are available, and this study illustrates a method of obtaining rock-fabric numbers and petrophysical class from stratigraphic descriptions for calculating permeability from the global transform.

ACKNOWLEDGMENTS

We thank James W. Jennings, Jr., Florence Bonnaffé, and Stephen C. Ruppel for their constructive comments on the manuscript. Lana Dieterich edited the manuscript. The illustrations were prepared with the assistance of the graphics group at the Bureau of Economic Geology (BEG). The support of this research by the Landmark Graphics Corporation via the Landmark University Grant Program is acknowledged. The publication is authorized by the director of BEG.

REFERENCES CITED

Harrington, R. R., and F. J. Lucia, 2012, Integration of rock fabrics and stratigraphy for petrophysical quantification of reservoir framework, *in* S. C. Ruppel, ed., Anatomy of a giant carbonate reservoir: Fullerton Clear Fork (Lower Permian) field, Permian Basin, Texas: Studies in Geology 63, p. 49–66.

Jennings Jr., J. W., and F. J. Lucia, 2003, Predicting permeability from well logs in carbonates with a link to geology for interwell permeability mapping: Society of Petroleum Engineers Reservoir Evaluation and Engineering, v. 6, no. 4, p. 215–225.

Jennings Jr., J. W., S. C. Ruppel, and W. B. Ward, 2000, Geostatistical analysis of permeability data and modeling of fluid-flow effect in carbonate outcrops: Society of Petroleum Engineers Reservoir Evaluation and Engineering, v. 3, no. 4, p. 292–303.

Kane, J. A., and J. W. Jennings Jr., 2012, Developing a wireline-log database and a basis for determining reservoir porosity, *in* S. C. Ruppel, ed., Anatomy of a giant carbonate reservoir: Fullerton Clear Fork (Lower Permian) field, Permian Basin, Texas: Studies in Geology 63, p. 67–76.

Lucia, F. J., 1995, Rock-fabric/petrophysical classification of carbonate pore space for reservoir characterization: AAPG Bulletin, v. 79, no. 9, p. 1275–1300.

Lucia, F. J., 1999, Carbonate reservoir characterization: New York, Springer-Verlag, 226 p.

Lucia, F. J., J. W. Jennings, Jr., M. A. Rahnis, and F. O. Meyer, 2001, Permeability and rock fabric from wireline logs, Arab-D reservoir, Ghawar field, Saudi Arabia: GeoArabia, v. 6, no. 4, p. 619–646.

Ruppel, S. C., and R. R. Harrington, 2012, Facies and sequence stratigraphy: Critical tools for reservoir framework definition, Fullerton Clear Fork Reservoir, Texas, *in* S. C. Ruppel, ed., Anatomy of a giant carbonate reservoir: Fullerton Clear Fork (Lower Permian) field, Permian Basin, Texas: Studies in Geology 63, p. 5–48.

Thomeer, J. H. M., 1960, Introduction of a pore geometrical factor defined by the capillary-pressure curve: American Institute of Mining, Metallurgical, and Petroleum Engineers Transactions, v. 219, p. 354–358.

Wang, F. P., and F. J. Lucia, 2012, Reservoir modeling and simulation of the Fullerton Clear Fork reservoir, Andrews County, Texas, *in* S. C. Ruppel, ed., Anatomy of a giant carbonate reservoir: Fullerton Clear Fork (Lower Permian) field, Permian Basin, Texas: Studies in Geology 63, p. 111–143.

5

Zeng, H., 2012, Construction and analysis of three-dimensional seismic porosity inversion models, *in* S. C. Ruppel, ed., Anatomy of a giant carbonate reservoir: Fullerton Clear Fork (Lower Permian) field, Permian Basin, Texas: Studies in Geology 63, p. 93–110.

Construction and Analysis of Three-Dimensional Seismic Porosity Inversion Models

Hongliu Zeng

Bureau of Economic Geology, The University of Texas at Austin, Austin, Texas, U.S.A.

ABSTRACT

Geology-guided reconditioning of seismic data is the key to improving extraction of relevant geologic information from three-dimensional (3-D) data volumes. In the Fullerton Clear Fork reservoir, the most convenient and useful tools for data reconditioning are phase shifting and high-frequency enhancement. A simple seismic phase rotation (to 90°) reconditions seismic data for impedance representation, roughly linking seismic amplitude directly to log lithology and porosity and making stratigraphic correlation more accurate. High-frequency enhancement raises the dominant frequency of 3-D seismic data from 30 to 50 Hz, improving seismic resolution without significantly deteriorating signal-to-noise ratio. More accurate and finer scale seismic mapping of reservoir parameters is achieved from model-based progressive inversion that seamlessly integrates geologic interpretation of well-derived seismic data and model-based seismic inversion for high-resolution (2 ft [0.6 m]) impedance models.

INTRODUCTION

Poststack seismic data contain rich information about reservoir stratigraphy, sedimentology, and physical properties. The value of seismic interpretation and well-seismic integration has long been recognized. Higher purposes of interpretation are (1) to establish a direct link between subsurface geology and the seismic signal by making seismic expression resemble the geologic section (structure, stratigraphy, lithology, rock properties, etc.) and (2) to extract the highest possible resolution information by seismic processing and by integration of well data. However, current seismic interpretation strategies fall short of these higher purposes by emphasizing structural mapping at the expense of stratigraphic evaluation. Stratigraphic information in seismic data is far from fully used because of the lack of thorough understanding of the relationship between stratigraphy and its seismic response, evidenced by regularly poorly conditioned seismic data that fail to express stratigraphy in a way that geologists can understand. The goal of this study is to develop better methods of stratigraphy-oriented data reconditioning and integration.

DOI:10.1306/13341540St633003

In this chapter, I describe a systematic approach to improving seismic stratigraphic interpretation. The first step in this approach is basic data reconditioning within the seismic-frequency range (5–70 Hz) to improve seismic data interpretability. Two key techniques were applied: phase rotation and high-frequency enhancement. A simple seismic phase rotation (to 90°) reconditions seismic data for impedance representation, roughly linking seismic amplitude directly to log lithology and porosity. High-frequency enhancement raises the dominant frequency of 3-D seismic data from 30 to 50 Hz, significantly improving seismic resolution.

However, a limit to the improvements that can be achieved in seismic data interpretability by basic data reconditioning exists. For example, 90° phasing does not eliminate the wavelet effect, and for wavelet side lobes to be removed for a truer representation of the geology, a seismic inversion must also be performed. Furthermore, limits to the resolution obtainable from the high-frequency enhancement exist. High-frequency enhancement processing does not result in a resolution high enough for reservoir-flow model construction. To reach such levels of resolution, well data are needed, and inversion without well-data constraining will not provide reservoir details beyond seismic resolution. To solve these problems, I applied a more advanced data integration method known as progressive inversion. With this approach, which is described in this chapter, a more accurate log-resolution (2 ft [0.6 m]) seismic mapping of reservoir parameters can be achieved.

METHODOLOGY

I emphasize stratigraphy-oriented 3-D seismic data interpretation by addressing three techniques.

Seismic Phase Adjustment to Layer-Based Event Expression

Standard seismic processing commonly produces (or at least attempts to produce) a 0°-phase seismic data as the final product. The interpretive advantages of 0°-phase data include wavelet symmetry, center lobe (maximum amplitude) coincidence with reflection interface, and higher resolution (Brown, 1991). However, these advantages are realized only if the seismic reflection comes from a single reflection interface. More commonly, one has to deal with seismically thin beds in which reflection amplitudes are composite seismic responses from the top and the base. In this situation, the observed waveform is a peak-trough couplet. Because seismic events do not coincide necessarily with bed geometry, tying lithology-porosity logs in wells to seismic data can be difficult, especially when multiple thin beds coexist in a stratigraphic interval. Ninety-degree–phase data are more suitable to geologic interpretation of seismically thin beds (Zeng et al., 1996, 2003; Zeng, 2003; Zeng and Hentz, 2004). The use of a 90°-phase wavelet overcomes the shortcomings of the 0°-phase wavelet by shifting the main lobe (maximum amplitude) of the seismic response to the center of the thin bed. Seismic response is symmetrical to the thin bed and not to the top and base of the bed, thus making the main seismic event (a trough, in this case) coincide with geologically defined high-frequency sequences (HFSs) and flow units. As a result, seismic polarity and/or amplitude is uniquely tied to stratigraphy. These improvements, when applied to real data, can make seismic interpretation easier.

High-Frequency Enhancement

By relatively suppressing low-frequency components and enhancing high-frequency components in the data, seismic resolution can be improved further. However, the processing will inevitably reduce the signal-to-noise ratio of the data. As a result, arguments still exist on how useful the technique is. Our experience indicates that the technique is valuable in enhancing thin-bed imagery without overboosting high-frequency noise beyond effective signal bandwidth. Several methods can achieve high-frequency signal enhancement, such as frequency filtering and deconvolution (Yilmaz, 1987) and spectral balancing (Tufekcic et al., 1981). For this study, spectral balancing was used to relatively enhance high-frequency energy in the data without overwhitening the spectrum.

Combination of Well and Seismic Data for Full-Spectrum Reservoir Modeling

Well-data integration is needed so that a high-resolution reservoir simulation model can be built from seismic data. The difficulty is on how to calibrate data of different scales and different resolutions in both geologically and computationally reasonable, seamless ways. Departing from the current trend of concentrating on geophysical algorithms, I emphasize the function of geologists in interpretation-oriented processing, such as seismic inversion. With the close involvement of geologists, integration (in the form of inversion) is typically not a one-time process. Zeng et al. (2003) and Tinker et al. (2004) demonstrated that a progressive (or repeated) inversion can improve the quality of reservoir modeling considerably. By upgrading the initial model from previous inversions and geologic knowledge of subsequent inversions, progressive inversions can combine and calibrate well and seismic data more efficiently and accurately.

DATA AND TIE

The data for this study come from Fullerton Clear Fork field in Andrews County, Texas (Figure 1). The Fullerton Clear Fork reservoir is composed of shallow-water carbonates of the Lower Permian (Leonardian) at a depth of about 6600 to 7500 ft (~2010–2280 m).

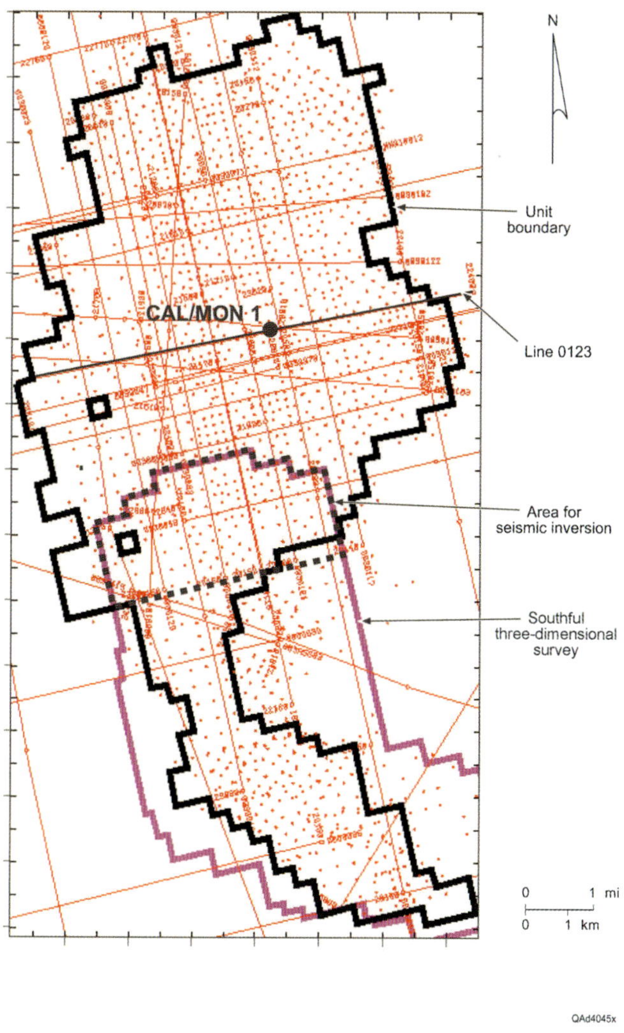

FIGURE 1. Location of well and seismic data and three-dimensional seismic inversion area in Fullerton field, Andrews County, Texas. CAL/MON 1 = well.

Available data for the field study include wireline logs (±1400 wells), cores (29, totaling ~15,000 ft [~4572 m]), and 3-D seismic (26 mi^2 [67 km^2]) and two-dimensional (2-D) seismic data (30 lines). The study focused on the central part of the field and used about 5 mi^2 (~13 km^2) of 3-D seismic and 66 well-log suites.

Well-to-seismic ties using synthetics are straightforward at Fullerton field. Figure 2 illustrates a seismogram generated using a sonic log from one of the wells in the field. The seismogram ties nicely to a 2-D seismic line across the well. Using well-log interpretation, one can easily pick Tubb and Abo reflectors from the seismic data, which form the top and base of the Lower Clear Fork and Wichita reservoir sections, respectively (see Ruppel and Harrington, 2012, for details on the stratigraphic units in the Fullerton Clear Fork reservoir). Synthetics in other wells in the field (both in 2-D and 3-D seismic areas) exhibit similar results.

Petrophysical analysis of wireline logs shows a linear relationship between acoustic impedance (AI) and calculated porosity. Although a spectrum of limestones, dolostones (from grain-dominated to mud-dominated dolostones), and anhydrite has been observed in conventional cores (Harrington and Lucia, 2012; Ruppel and Harrington, 2012), log AI steadily decreases with log porosity in a depth range of 6600 to 7300 ft (2010–2220 m) at a relatively high level of correlation ($\rho = 0.88$) (Figure 3). This close relationship between AI and porosity implies that seismic amplitude should be a good indicator of porosity, provided that amplitude-tuning effects can be reasonably estimated using well control.

However, the observed excellent well log-to-seismic tie and the simple acoustic relationship do not automatically lead to an easy interpretation of seismic data, especially in terms of stratigraphy and porosity. Figure 4 shows a standard 0°-phase seismic section tied to well logs. Note that only major stratigraphic boundaries (Tubb, Wichita, and Abo; solid lines in Figure 4) are linked to prominent seismic events. High-frequency sequence boundaries (HFS 2.1–HFS 2.3, Wichita 8; Figure 4) are not resolved by seismic and cannot be correlated on the basis of seismic phase. Low-AI or high-porosity zones in HFSs (dashed lines, Figure 4) recognized in sonic and/or density and porosity logs tie neither to polarity nor to amplitude. A similar situation is seen in other seismic data volumes (2-D and 3-D) in the project area. This observation suggests that, in their original state, the seismic data are not in the best form for

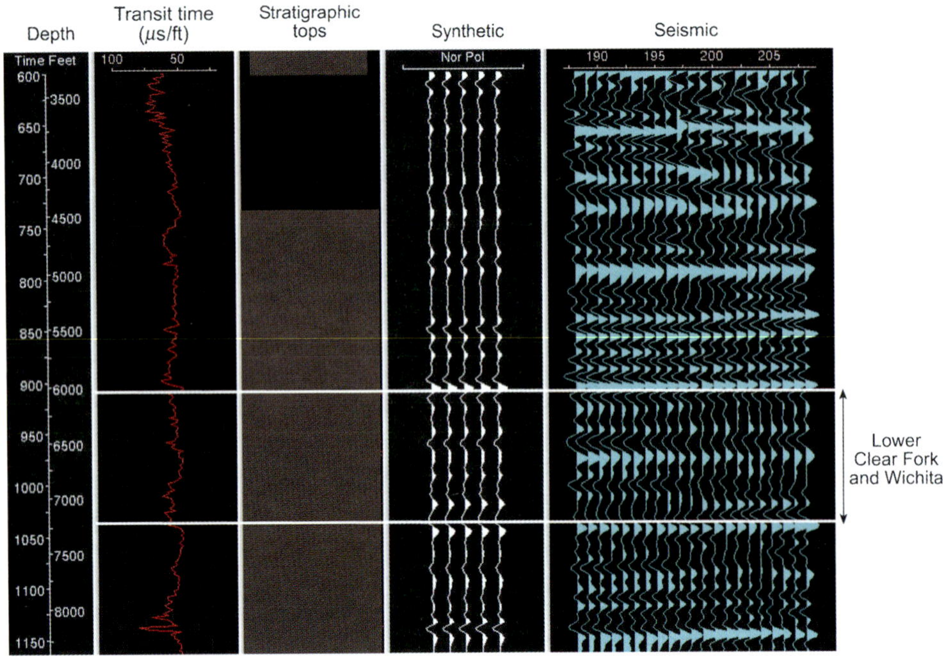

FIGURE 2. Synthetic seismogram of Lower Clear Fork and Wichita and adjacent formations.

stratigraphic interpretation and must be reconditioned and reprocessed before being used for this purpose.

IMPROVEMENT OF SEISMIC DATA INTERPRETABILITY

Simple poststack processing tools are available in various seismic interpretation packages (e.g., Poststack in Landmark). A quick improvement of data interpretability can be achieved by applying appropriate data processing techniques for better stratigraphically oriented seismic interpretation. My tests show that seismic phase adjustment and high-frequency enhancement are two of the most useful tools.

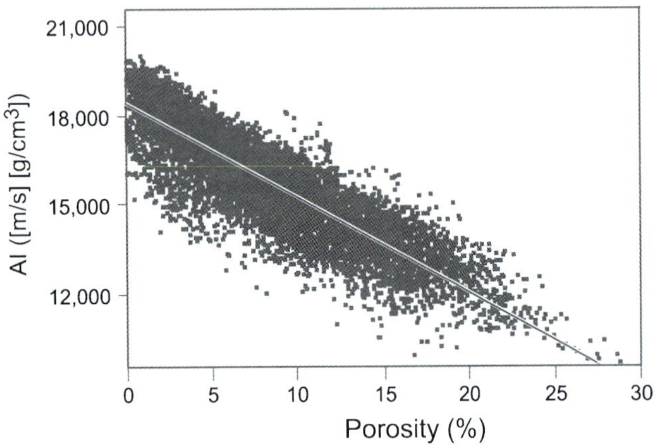

FIGURE 3. Linear relationship between log acoustic impedance (AI) and log porosity in the Lower Clear Fork and Wichita, Fullerton field, Andrews County, Texas. Acoustic impedance was computed from transit time and bulk density logs in eight wells, whereas porosity was calculated by neutron-density crossplotting, with a sample rate of 0.5 ft (0.2 m). Correlation coefficient between AI and porosity = 0.88.

Ninety-Degree Phasing

As seen in Figure 4, standard 0°-phase seismic data are not optimal for stratigraphically oriented thin-bed interpretation. To improve data interpretability, the 0°-phase seismic data in Figure 4 were reprocessed by applying a 90°-phase shift (Figure 5). In this reprocessed seismic section, the important stratigraphic boundaries (Tubb, Wichita, and Abo) are tied to seismic events with a 90°-phase shift (zero crossings, Figure 5). Although HFS boundaries still fail to correlate to specific seismic phase characters, high-porosity zones (low AI) associated with some of them (dashed lines, Figure 5) tend to tie to seismic trough events (negative amplitude). The resolution, however, is dominated by the low dominant frequency of the data (~30 Hz), which causes some thinner porosity zones to be imaged as merged seismic events without clear separation.

Apparently, the 90° phasing of seismic data reconditions seismic traces to better resemble an impedance profile so that interpretability of the data is improved. The procedure is simple, reliable, and not data damaging.

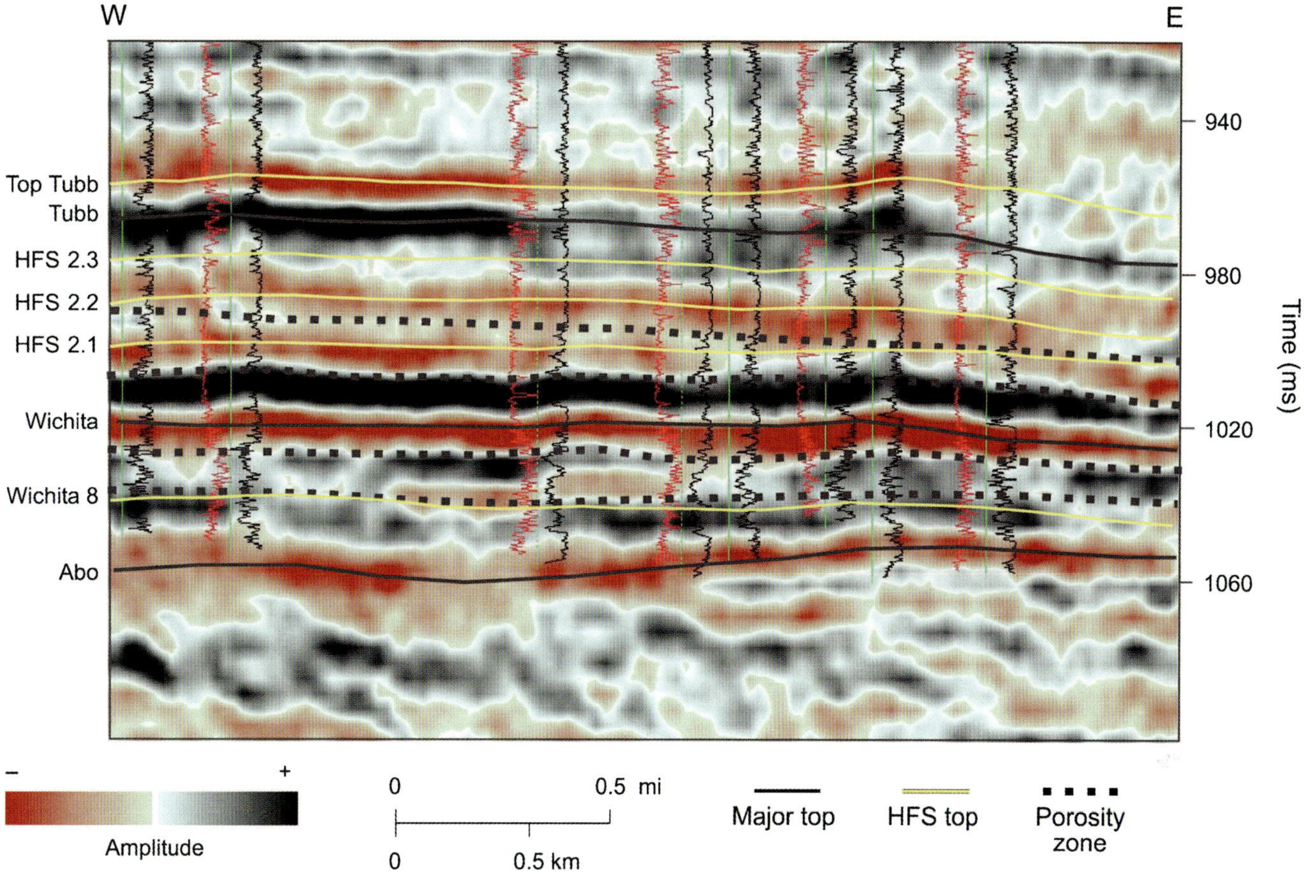

FIGURE 4. A 0°-phase seismic section from a three-dimensional survey area showing a poor tie between high-porosity or low-AI units and amplitude or polarity and difficulty in picking high-frequency sequence (HFS) boundaries. The dashed lines highlight porosity zones in HFSs and flow units. Log in red = gamma ray; Log in black = porosity.

High-Frequency Signal Enhancement

Although 90° phasing significantly improves visual correlation between seismic events and stratigraphic units (Figure 5), the resolution of seismic data is still not satisfactory. It is still difficult to map individual thin porosity zones in HFSs because of the common merging of seismic events following these closely spaced units. To improve seismic resolution, spectrum-balancing processing was applied to the data (Figure 6), after which the dominant frequency of 3-D seismic data was raised from about 30 to 50 Hz. As a result, seismic events are shorter in time and are better separated on seismic sections. Eventually, seismic separation between thin porosity zones becomes clearer (cf. Figures 5, 6).

Although high-frequency enhancement processing raises the noise level in the data to a certain degree so that the signal-to-noise ratio of the data is somewhat reduced, the new seismic data set provides a better definition of major reservoir units in the field. With this increased resolution, structural mapping of major stratigraphic boundaries (Tubb, Wichita, and Abo) and attribute extraction of thin porosity zones in HFSs can be performed in the entire study area. However, seismic resolution is still not enough for seismic correlation of HFS boundaries and flow-unit-level reservoir units because of the lack of seismic information beyond 70 Hz.

WELL-BASED ACOUSTIC IMPEDANCE MODEL

To further improve reservoir interpretation and modeling, well data must be integrated with seismic data. Cores and wireline logs provide not only measurements of lithology, facies, and rock properties, but also high-resolution (down to 0.5 ft [0.2 m]) correlation of geologic and flow units that is beyond seismic control. It is important to use well information to the fullest extent possible. At the current stage of seismic reservoir characterization, to guide data integration (inversion), the best possible high-frequency sequence-stratigraphic and high-resolution AI models need to be constructed from well data.

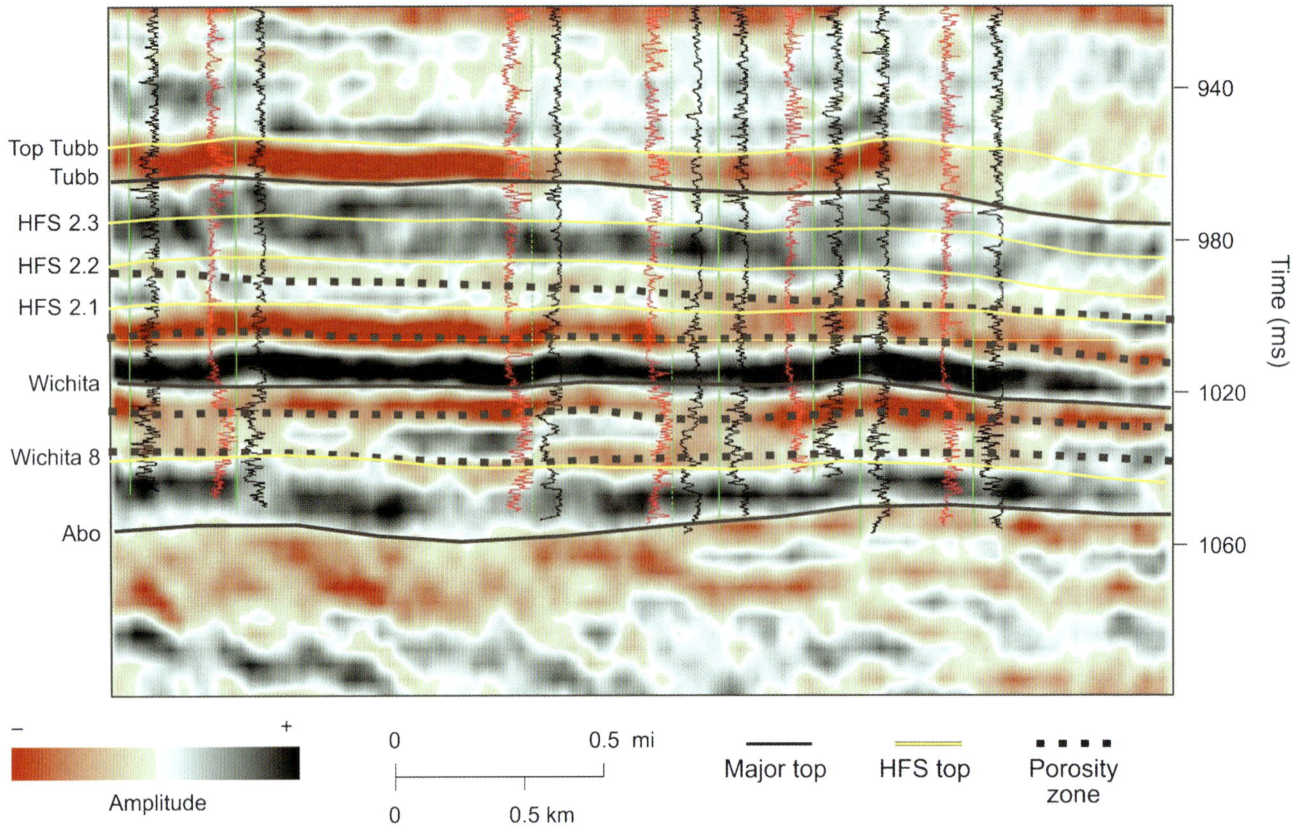

FIGURE 5. A 90°-phase seismic section in a three-dimensional survey area showing an improved tie between high-porosity or low-AI units and amplitude or polarity. The identification of high-frequency sequence (HFS) boundaries, however, is still difficult.

Well-log-based AI models can be divided into four basic types on the basis of construction methods and output model resolution (Figure 7):

1) Background model (Figure 7A). For this model, only a low-frequency AI trend is calculated from well AI logs. No stratigraphic surfaces are used.
2) Blocky model (Figure 7B). This model incorporates stratigraphic surfaces including high-frequency sequence-stratigraphic boundaries. Acoustic impedance values are averaged for each layer defined by stratigraphic horizons picked from well and seismic data, producing blocky impedance patterns. In this model, horizons are commonly seen as sharp AI boundaries. The resolution depends on the number and spacing of stratigraphic surfaces (horizons) and is typically lower than seismic resolution.
3) Smoothed log model (Figure 7C). Acoustic impedance values between wells are based on filtered well-log trends within the interpreted stratigraphic framework. The resolution is variable. In this case, model resolution is adjusted to seismic resolution (<70 Hz).
4) High-resolution log model (Figure 7D). Acoustic impedance values are modeled at wireline-log resolution (~1.25 ft [~0.38 m]). Thin beds between wells are linearly interpolated between successive pairs of horizons. As a result, modeled reservoir geometry and architecture are very sensitive to the stratigraphic framework provided.

These models (Figure 7A–D) use progressively detailed well-data information. The degree of the involvement of a geologist, through interpretation of stratigraphic framework and reservoir architecture used in the models, also increases. For best possible integration, it is important to construct and use well-based AI models at a log-resolution level.

MODEL-BASED SEISMIC INVERSION

Achieving integration of well and seismic data in reservoir modeling is equivalent in geophysical terms to completing three tasks: removing the wavelet effect, restoring the low-frequency AI trend, and combining high-frequency (high-resolution) well AI information.

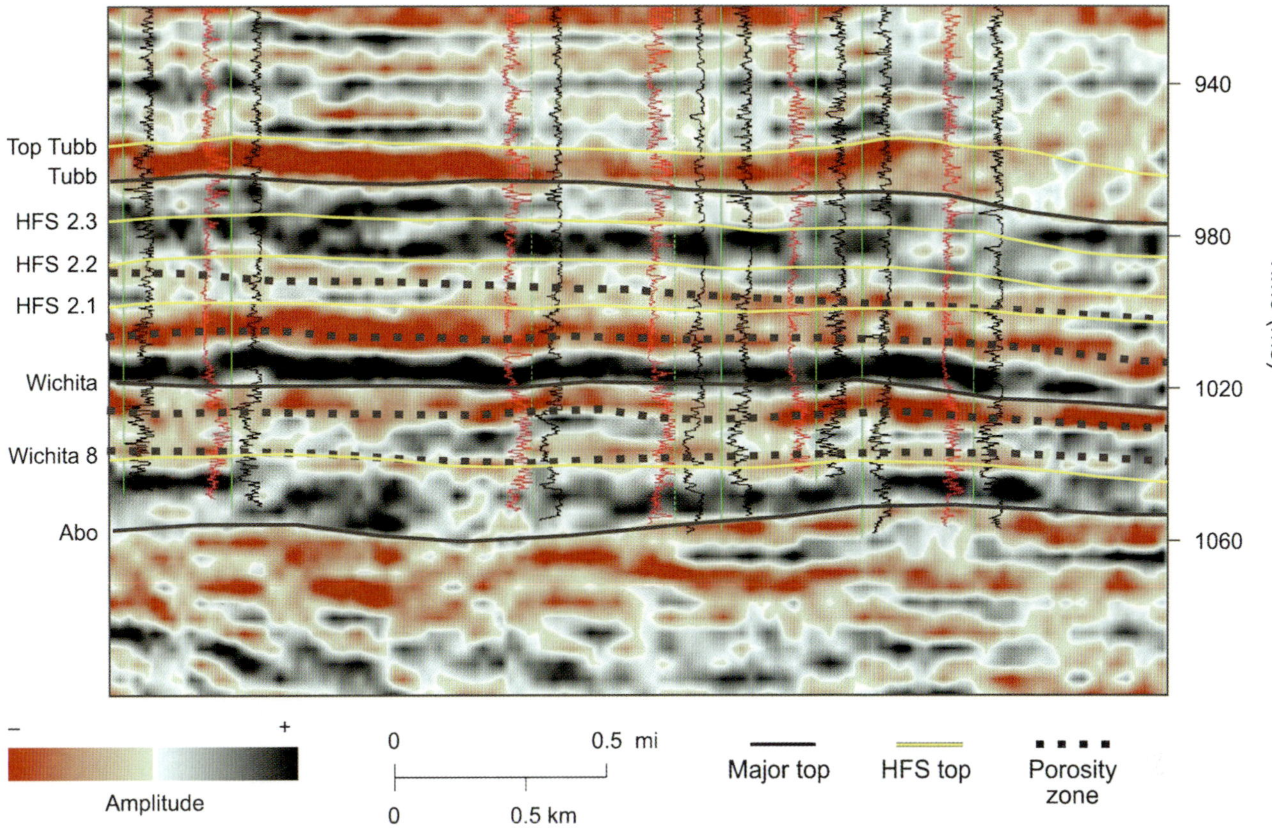

FIGURE 6. High-frequency enhanced (spectral-balanced) version of Figure 5 showing an improvement in seismic resolution. Although not precise, the position of high-frequency sequence (HFS) boundaries can be approximated at the tops and bottoms of the porosity zones.

Appropriate seismic inversion can accomplish all three seamlessly and efficiently.

Why Model Based?

Geophysical literature names four main types of inversion:

1) Recursive inversion. Assuming a seismic trace to be a reflection-coefficient series, AI can be calculated by recursively solving the reflection-coefficient equation. In this process, the wavelet effect is not removed, and well data are not involved except to provide a low-frequency trend. Results are commonly poor because of a large prediction error.
2) Sparse spike inversion. This method finds major reflection interfaces (spikes) that make a significant contribution to a seismic trace by inserting reflection interfaces at proper positions for the best fit of synthetics and real seismic traces. Minor reflection interfaces are ignored. As a result, inversion is low frequency in nature, and the wavelet effect is reduced. Detailed well data, however, cannot be combined into output in this process.
3) Neural network inversion. An artificial neural network can be trained to find nonlinear correlations between seismic attributes and AI (or an appropriate rock property of interest). When this process is used, the wavelet effect can be reduced by forced calibration of seismic traces to wireline logs. The resolution, however, is still limited to the seismic-frequency range.
4) Model-based inversion. Also called generalized linear inversion, this approach requires the input of an initial (or guess) AI model. The initial model is typically constructed from available well and seismic data. If this model is built such that the full spectrum (low frequency, seismic frequency, and high frequency) of information is included, the inversion process will modify the model at the seismic-frequency range by minimizing error functions between synthetics and seismic traces and let the low-frequency trend and high-frequency details from well interpretation carry through to the final model. Model-based inversion is the most popular inversion package in the industry.

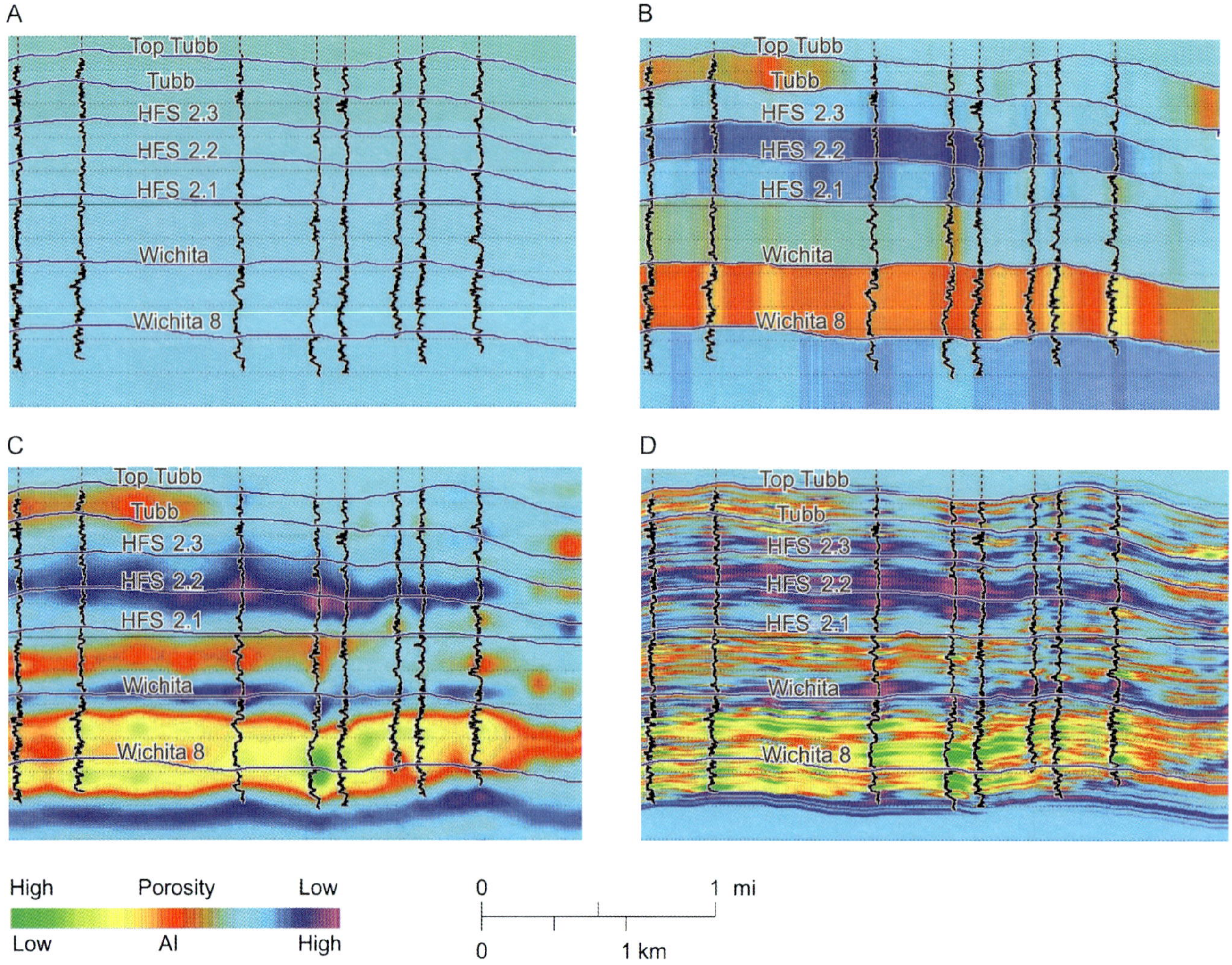

FIGURE 7. Wireline-log-based acoustic impedance (AI) models for reservoir characterization and seismic inversion. A total of seven high-frequency sequence (HFS) boundaries were used for stratigraphic control of models. (A) Background model. (B) Blocky model. (C) Smoothed log model at seismic frequencies (0–70 Hz). (D) High-resolution log model sampled at 3 to 3.9 ft (1–1.2 m) (0.125 ms). See text for discussion. The AI is calculated from porosity logs. Note how the resolution increases from models A to D and how the log sample frequency increases.

Currently, only model-based inversion offers a platform for building a full-spectrum reservoir model in a single process that combines all available information. The model-based inversion method is the approach used in the Fullerton Clear Fork reservoir study.

How Well and Seismic Data Are Integrated

A model-based inversion is typically an iterative process (Figure 8) of three steps:

1) Construct an initial AI model trace.
2) Create synthetics using the estimated AI trace and extracted wavelet.
3) Calculate the difference between the synthetics and the real seismic trace. If the error is within the tolerance level, output the AI trace as an inversion model; otherwise, compute the objective function and model perturbation, update the AI trace, and go back to step 1.

Because the initial model is seldom perfect, the process is commonly repeated multiple times.

In the best-case scenario, a model-generated synthetic seismogram fits with its real seismic counterpart. The process is restricted to the seismic-frequency range in the real data. Any frequency components beyond seismic in the initial model (both low and high frequencies) are filtered out in the synthetics by the extracted wavelet and cannot influence optimization. If restraints to output AI are set loosely enough, model components beyond the seismic-frequency range

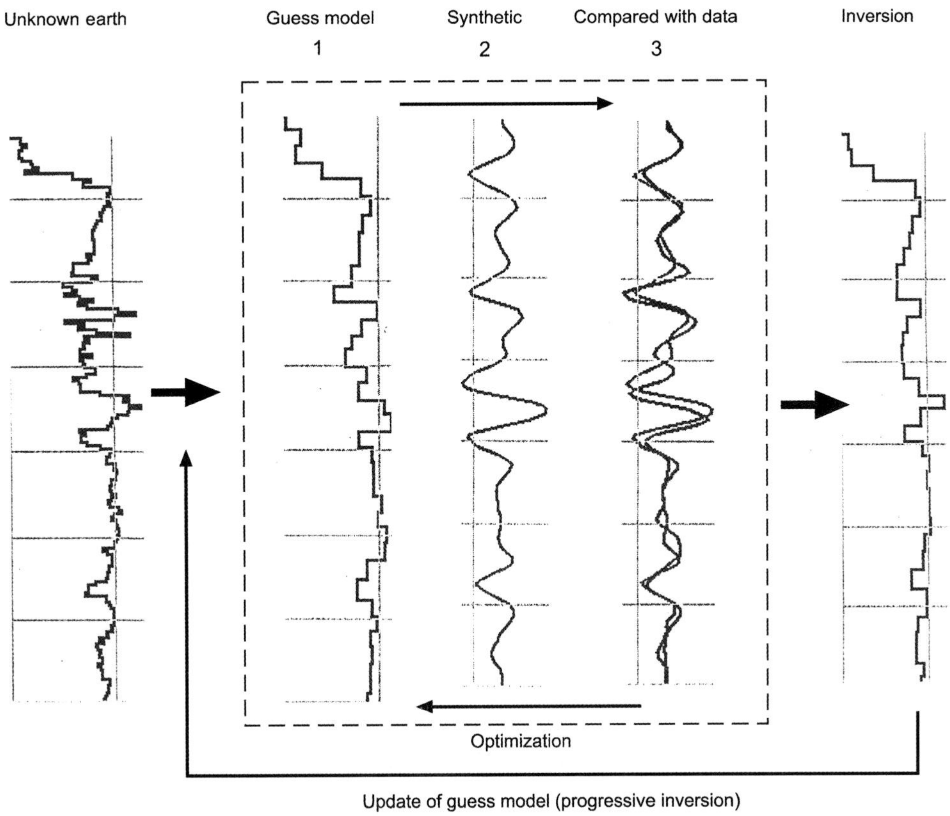

FIGURE 8. Steps in model-based inversion.

tend to carry through to the updated AI models and the final inversion.

As an example, Figure 9 shows four versions of inversion results from the four different initial well-log models in Figure 7. Although the same well data were used in the well-log AI initial models (Figure 7) and the inversions (Figure 9), differences between the two can be quite significant. Because the background AI model (Figure 7A) has no stratigraphic control, the inversion based on this model is controlled mainly by seismic frequency (Figure 9A). A better balance between log AI and seismic is achieved in the blocky AI model inversion (Figure 7B) that is constrained by high-frequency sequence-scale horizons. However, this inversion (Figure 9B) emphasizes horizontal continuity at and near the sharp impedance boundaries at these geologic boundaries and can be biased. The smoothed log model (Figure 7C) better follows the AI trend defined by well logs and therefore better honors geologic knowledge (both stratigraphic framework and AI distribution). The inversion, however, still lacks detail beyond the seismic-frequency range. The differences in inverted AI for the first three inversions are shown in Figure 10. In some intervals (e.g., intervals between HFS 2.2 and HFS 2.3 and between Wichita and Wichita 8), a 25% variation in inverted AI is observed among the three inversions, although all inversions fit seismic data equally well. Finally, introduction of high-frequency cycle-scale log data to the initial model (Figure 7D) leads to a high-resolution inversion (Figure 9D). The similarity between the model and the inversion is a result of the carry-through of high-frequency AI components from the model to the inversion. The inversion, however, still incorporates a seismic signal at the seismic-frequency range, which modifies size, continuity, and AI value for many of the thin reservoir units in the initial model.

PROGRESSIVE INVERSION

Problems in Model-Based Inversion

Assuming good-quality well and seismic data and correct implementation of inversion software, many factors can still influence the quality of inversion because of the nonuniqueness of inverse problems. Geophysicists tend to concentrate more on improving optimization algorithms. Vigorous model testing indicates that, when a good geologically constrained model is used, a regular optimization algorithm should

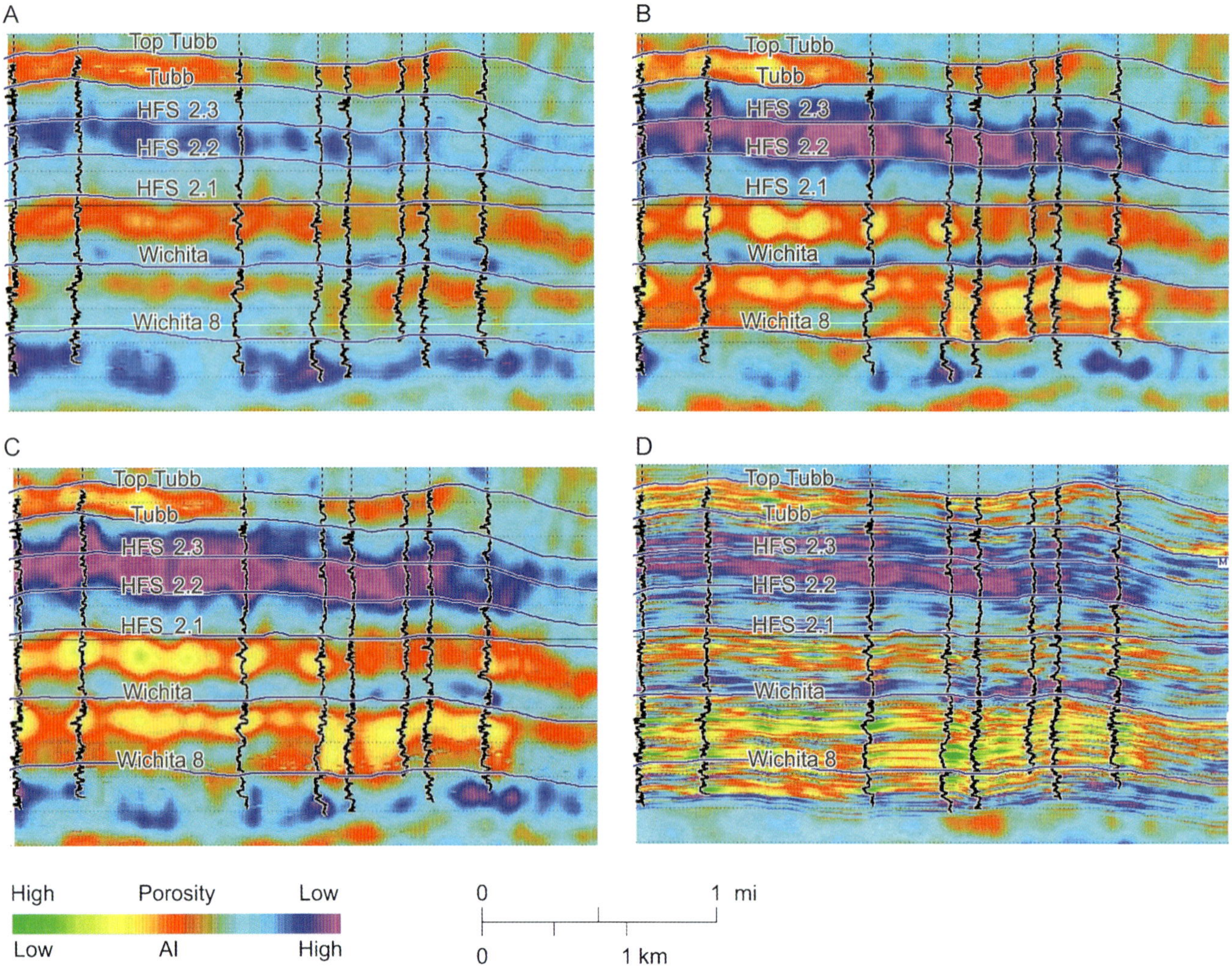

FIGURE 9. Seismic inversions of well-log acoustic impedance (AI) models shown in Figure 7. Note that the inversion resolution increases with increasing stratigraphic control in models. HFS = high-frequency sequence.

yield satisfactory results. However, model quality is equally important. A bad model will lead to a bad inversion, even if the most sophisticated optimization algorithm is applied. In this study, model quality is our major concern.

From the geologist's point of view, a high-quality initial model is created by deriving the most detailed, most accurate stratigraphic framework possible from available well and seismic interpretation and then spatially populating AI within the framework in a geologically and statistically reasonable way. In this study, accuracy of the stratigraphic framework was the key. Although varying approaches to interwell mapping are important factors in model construction, a discussion of attribute mapping is beyond the scope of this study. All well data were mapped between and beyond wells, following the inverse-square distance rule, a popular technique in the industry.

In the final analysis, the task of performing a good model-based inversion is reduced to a surface-picking issue. The biggest challenge is that it is difficult, if not impossible, to build an accurate high-resolution initial model by interpreting closely spaced stratigraphic surfaces between and beyond wells using low-resolution seismic data (Figures 4–6). In this seismic data volume, good horizons that can be at least approximately tied to geologic surfaces are limited (e.g., top Tubb, Tubb, and Wichita; Figure 11). However, these surfaces alone are not enough to accurately define the stratigraphic model. Picked geologic surfaces (e.g., Wichita; Figure 6) may offset seismic events (in this case, a peak-to-trough zero crossing) because of seismic interferences or misperception in interpretation. High-frequency and cycle boundaries are not resolved commonly and cannot be picked directly from low-frequency seismic sections (e.g.,

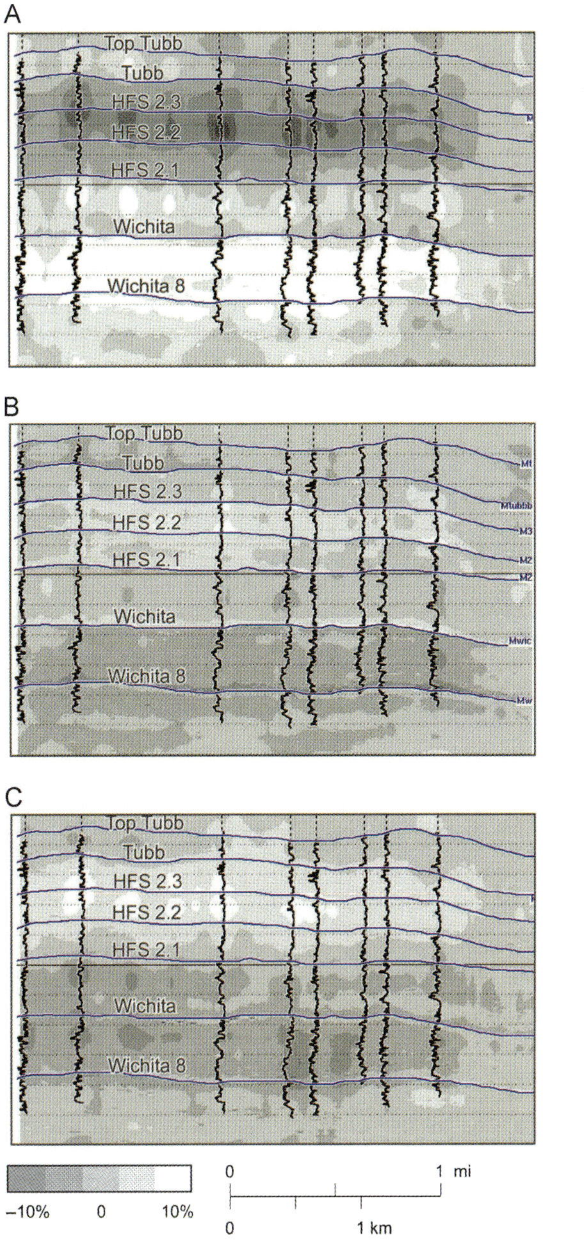

FIGURE 10. Differences of acoustic impedance inversion corresponding to different initial model inputs. (A) Difference between background model inversion (Figure 9A) and average value of models A, B, and C in Figure 9. (B) Difference between blocky model inversion (Figure 9B) and average value of models A, B, and C in Figure 9. (C) Difference between smoothed model inversion (Figure 9C) and average value of models A, B, and C in Figure 9. HFS = high-frequency sequence.

HFS 2.1, HFS 2.2, HFS 2.3, and Wichita 8; Figure 6). Without sufficient good-quality surfaces being mapped, the initial model is almost always oversmoothed and poor in quality. To make things worse, most people use this poor initial model once and only once in the inversion process and never revisit it. As a result, significant error and/or uncertainty in model building tends to carry through to the final inversion product.

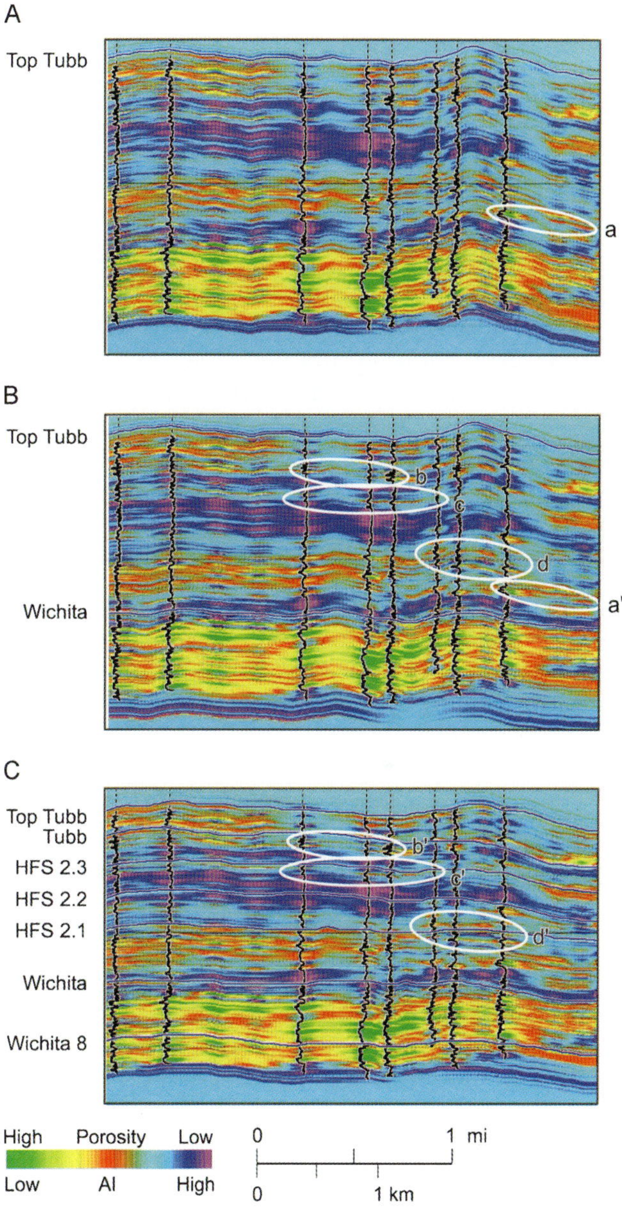

FIGURE 11. Initial wireline-log-based acoustic impedance (AI) models used for progressive inversion. (A) Model using one stratigraphic horizon, the top Tubb. (B) Model using two stratigraphic horizons, the top Tubb and the top Wichita. (C) Model based on seven horizons (as shown). The logs are pseudo-AI calculated from porosity. Miscorrelations of thin-bed units in early-generation models (a–d) are corrected in late-generation models (a′–d′, respectively) by incorporating more and more geologic control in wells. HFS = high-frequency sequence.

Proposed Solution

Progressive inversion improves model-based inversion by means of multiple progressive initial models being built and multiple inversions being performed (Figure 8). The first initial model should be a low-resolution model that is based on a small number of prominent and reliable geologic boundaries and seismic horizons. The first inversion is performed on the basis of this initial model, which reduces the wavelet effect and commonly reveals more geologic detail than the original seismic data. Horizons are then reinterpreted by comparing the seismic data and the first inversion. Additional horizons are then added through the interpretation of seismic data and first inversion to create a new and more accurate initial model that fits geologic expectations. A new inversion based on this improved initial model will provide even more accurate geologic details. This process may be repeated until the inversion is satisfactory for reservoir-model-building applications.

For this study, a progressive inversion was performed using the Hampson-Russell inversion software (the models are shown in Figure 11, and the inversions, in Figure 12). The first inversion was done using an initial impedance model (Figure 11A) made from a single seismic horizon, the top of the Tubb Formation, which is one of the most prominent geologic contacts in the wells and one that shows the best tie to the seismic data (Figure 6). Using only one horizon, I assumed a sheetlike formation without lateral thickness variation. The resulting inversion (Figure 12A), as expected, is less than desired; many of the thin reservoir units seen in the inverted section do not honor well correlations. However, compared with the seismic data (Figure 6), the inversion does a much better job of resolving other geologic boundaries. In subsequent rounds of modeling and inversion, progressively more stratigraphic surfaces were added to improve stratigraphic resolution (Figures 11B, C; 12B, C). The final inversion (Figure 12C) was performed on an initial model (Figure 11C) that incorporates all HFS boundaries recognizable from the interpretation of well data, seismic amplitude data, and intermediate inversion results.

A key improvement achieved through progressive inversion is the introduction of an increasingly more accurate stratigraphic framework via more and more horizons. Although not obvious from the visual inspection of Figures 11 and 12, the seven horizons used in the final-round inversion provide the most detailed stratigraphic correlation based on wireline-log, core, and seismic analyses. This is better seen in Figure 13, which shows the closeness of fit between well-log-defined stratigraphic tops and seismic events. With this level of inversion, thickness variations in each of the HFSs are easily modeled, which significantly improves the definition of reservoir architecture and flow units in the model. By contrast, the limited stratigraphic control horizons in the

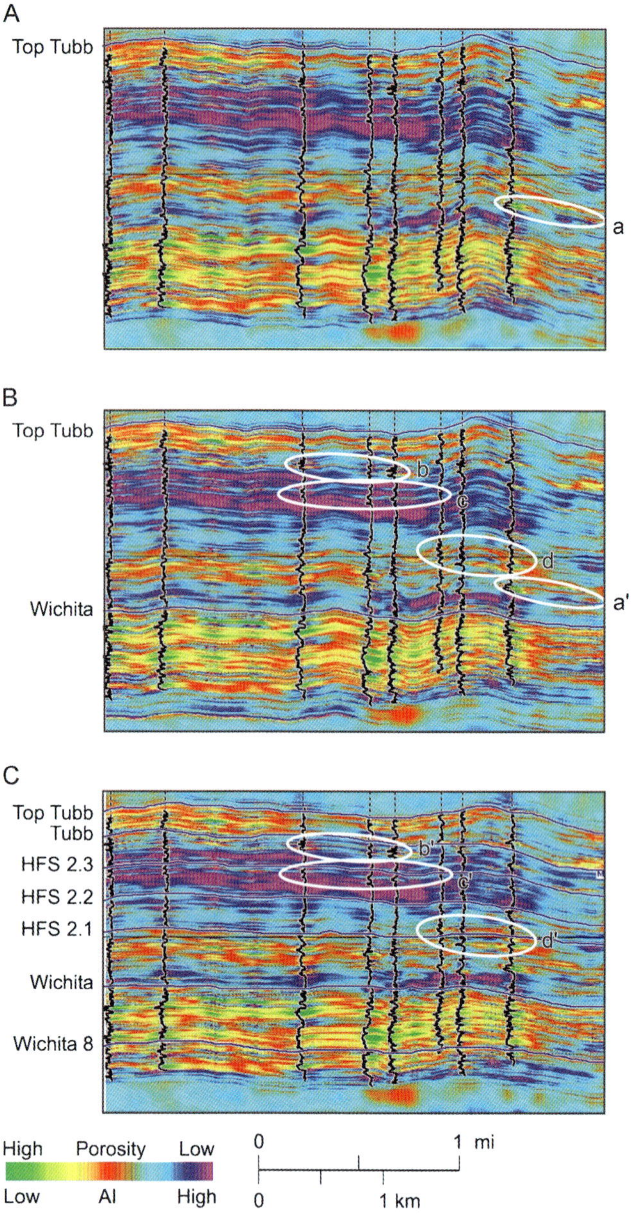

FIGURE 12. Progressive inversion. (A) Inverted acoustic impedance (AI) based on a one-horizon model (Figure 11A). (B) Inverted AI based on a two-horizon model (Figure 11B). (C) Inverted AI based on a seven-horizon model (Figure 11C). Miscorrelations of thin-bed units in early-generation inversions (a–d) are corrected in late-generation inversions (a′–d′, respectively) by incorporating more and more geologic control in wells. HFS = high-frequency sequence.

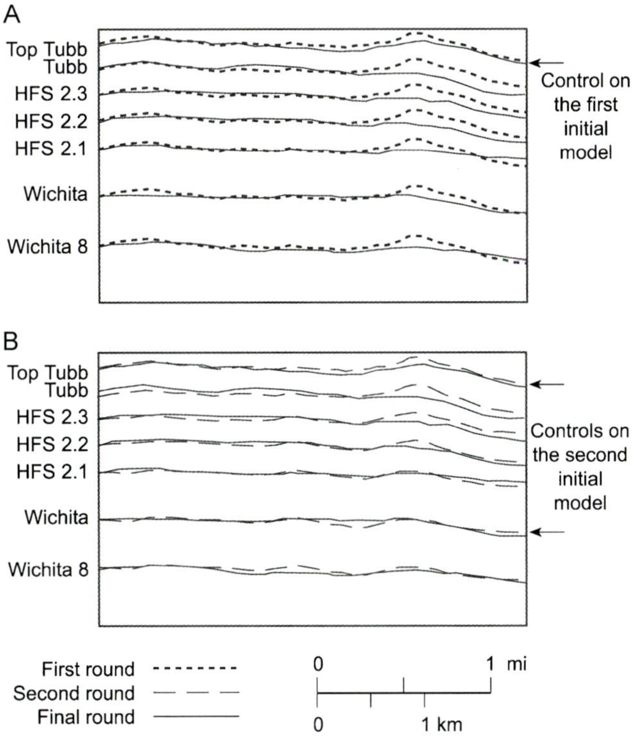

FIGURE 13. **Comparison of how the definition of stratigraphic surfaces in inversions improves with increasing numbers of horizons. (A) One-horizon model (first round) versus seven-horizon model (final round). In the one-horizon model, high-frequency sequence (HFS) boundaries were extrapolated parallel from the single control surface (top Tubb). (B) Two-horizon model (second round) versus seven-horizon model (final round). In the two-horizon model, HFS boundaries Tubb and HFS 2.1 to 2.3 were linearly interpolated between the control surface tops of the Tubb and the Wichita. High-frequency sequence boundary Wichita 8 was extrapolated parallel to the control surface of the top Wichita.**

first- and second-round inversions (Figures 11A, B; 12A, B; 13A, B) display correlation errors ranging from 0 to 50 ft (0–15 m), more than enough to produce a miscorrelation of flow units between wells. Some of the effects of these miscorrelations can be seen by comparing early-generation models with late-generation models (Figure 11a–a′, b–b′, c–c′, and d–d′). These miscorrelations are also apparent in the inversions (Figure 12a–a′, b–b′, c–c′, and d–d′). Miscorrelation can also be illustrated by AI differences in successive rounds of inversion (Figure 14). At or near the wells, all models generate similar results (white vertical strips in Figure 14A, B) because of the dominance of well-log control. Differences in AI estimation are far greater, however, in interwell areas. These differences are reduced by progressive rounds of inversion because of the improvements in thin-bed definition from fine-tuning geologic surface picking and integration of more geologic surfaces. Although all three inversions fit seismic data equally well, significant changes of AI distribution are apparent (high-frequency bands in Figure 14A, B). The final (fourth round) inversion should provide the best fit between geologic interpretation and seismic signal.

Progressive inversion is a seamless way to integrate geologic knowledge and seismic processing for reservoir characterization. It requires geologists to be intimately involved in the seismic inversion process. A sequence-stratigraphic model provides guidance on how seismic correlations should be conducted between and beyond wells for initial guess-model construction using well impedance logs. A comprehensive study of core, logs, and seismic data further

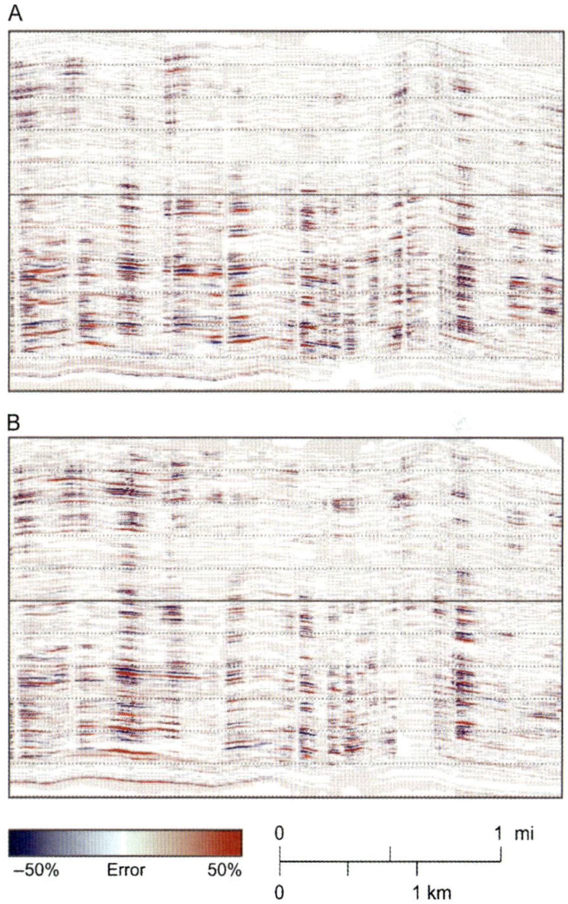

FIGURE 14. **Each step in progressive inversion changes high-frequency acoustic impedance (AI) layering and distribution, shown by differences in AI estimations through progressive inversion. (A) Difference between second-round inversion (Figure 12B) and first-round inversion (Figure 12A). (B) Difference between final-round inversion (Figure 12C) and second-round inversion (Figure 12B).**

established a high-frequency sequence-stratigraphic framework at all well locations (Ruppel, 2003; Ruppel and Harrington, 2012). In the study area, the geologic model indicates that all HFS-level seismic picks in the Lower Clear Fork and Wichita are relatively flat because of the deposition on a low-relief shallow-water platform (Ruppel and Harrington, 2012). Initial and intermediate inversion results allow correlation of HFS boundaries between and beyond well locations that honor both well picks and seismic trends. Multiple seismic and/or geologic picks in a short time window (in this case, seven horizons in 100 ms) transfer detailed a-priori stratigraphic information to seismic inversion, resulting in impedance models that exceed expectations if well or seismic data alone are used.

Blind Testing

So that the value and limits of seismic signal and progressive inversion could be evaluated, the seven-horizon initial AI model was rebuilt with nine wells removed from the 66-well database. A new inversion was then done following the same procedures and parameter setup. Comparisons of the two models and their respective inversions are displayed in Figures 15–17. In each figure, impedance distribution in a slice through the original model and inversion is compared with impedance distribution in the same slice through the model and inversion made after wells have been removed. Two major observations follow:

1) Seismic data can supply important missing geologic information between and beyond well control. The removal of one well (well 2) in Figure 15 significantly changes the well-log model (cf. a, b in Figure 15A, B). By contrast, the inversion models are very similar in both porosity distribution and flow-unit architecture (cf. a, b in Figure 15C, D). Figure 16 not only shows a similar impact of missing wells on the well model (cf. d in Figure 16A, B), but also shows that, in some cases, the inversion can also be significantly modified by missing wells (cf. d in Figure 16C, D). This kind of situation could be very common along the margins of fields or in areas of poor well control. In Figure 17, by contrast, the removal of well control resulted in only minor changes in the well-log model and in the inversions (cf. Figure 17A–D). This probably reflects good surrounding well control and agreement of well data and seismic data.
2) High-resolution details in the inversion are imperfect. Seismic signal is of low frequency, with a resolution no better than one-fourth wavelength (~66 ft [~20 m] in this case). The high-resolution layering (1 ft [0.3 m] in the initial model and 2 ft [0.6 m] in the inversion) is defined by well AI logs within a stratigraphic framework between and beyond wells. Seismic inversion adjusts AI distribution without changing the thickness architecture. Unfortunately, in this study, only the simplest attribute mapping was conducted between wells: thin-bed thickness is proportional to interval thickness between horizons; AI distribution in each bed is predicted by the inverse-square distance rule. As a result, the errors in thin-bed thickness and AI mapping are not trivial, leading to imperfect AI estimation. Inversions with or without missing wells, therefore, show somewhat different details near the well location (e.g., around area d in well 2 in Figure 16; around area c in well 3 in Figure 17). Some isolated thin beds are completely unrepresented without the use of the missing wells. Further improvements in the inversion could be achieved by applying more advanced mapping algorithms.

CONCLUSIONS

Poststack seismic data are commonly not in optimal form and require reconditioning before they can be used for stratigraphic analysis and reservoir characterization.

1) Basic data conditioning applies do-it-yourself poststack processing tools to improve data interpretability and resolution. A 90° phasing adjusts seismic traces to resemble impedance logs; a high-frequency enhancement enables interpreters to identify more and thinner geologic units (at a 100 ft [30 m] resolution level).
2) Progressive seismic inversion can provide a good representation of reservoir porosity in areas of missing or poor-quality log data.
3) Progressive seismic inversion requires the involvement of geologists by providing a geologic model and quality-checking seismic picking of high-resolution geologic boundaries. Progressive inversion promises to seamlessly integrate geologic knowledge and seismic data for detailed impedance modeling down to wireline-log resolution.

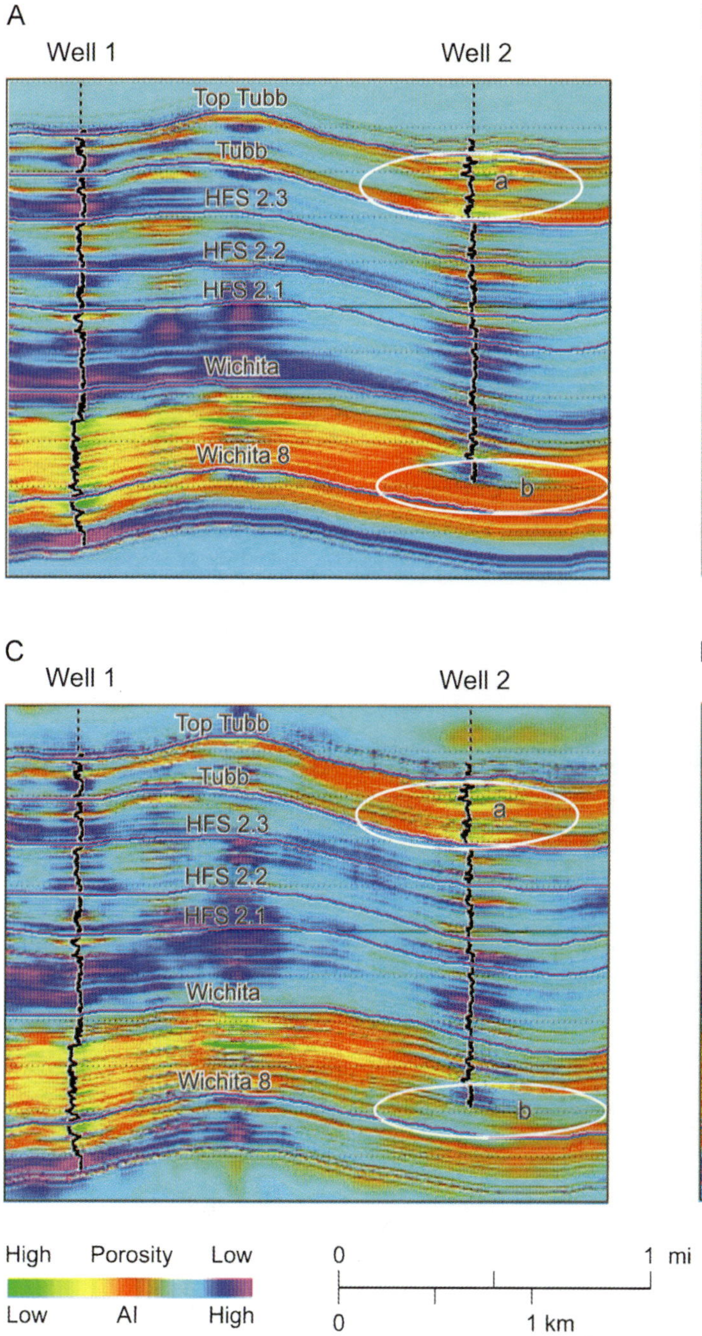

FIGURE 15. Blind test showing the value and the limit of progressive inversion (example 1). (A) Initial log-based acoustic impedance (AI) model. (B) Initial model with well 2 removed. (C) Seismic inversion of initial model in A. (D) Seismic inversion of initial model in B. Note that, when well 2 is removed from the model in A, a high-porosity zone (a) and a low-porosity zone (b) in A are absent in B. By contrast, the seismic inversion images these high- and low-porosity zones even without the missing well (cf. C and D). HFS = high-frequency sequence.

ACKNOWLEDGMENTS

The constructive critical review comments by Julia Gale and Stephen C. Ruppel greatly improved the manuscript. Lana Dieterich edited the manuscript. The illustrations were prepared with the assistance of the graphics group at the Bureau of Economic Geology (BEG). The support of this research by Hampson-Russell Software Services Ltd., and the Landmark Graphics Corporation via the Landmark University Grant Program are acknowledged. The publication is authorized by the director of BEG.

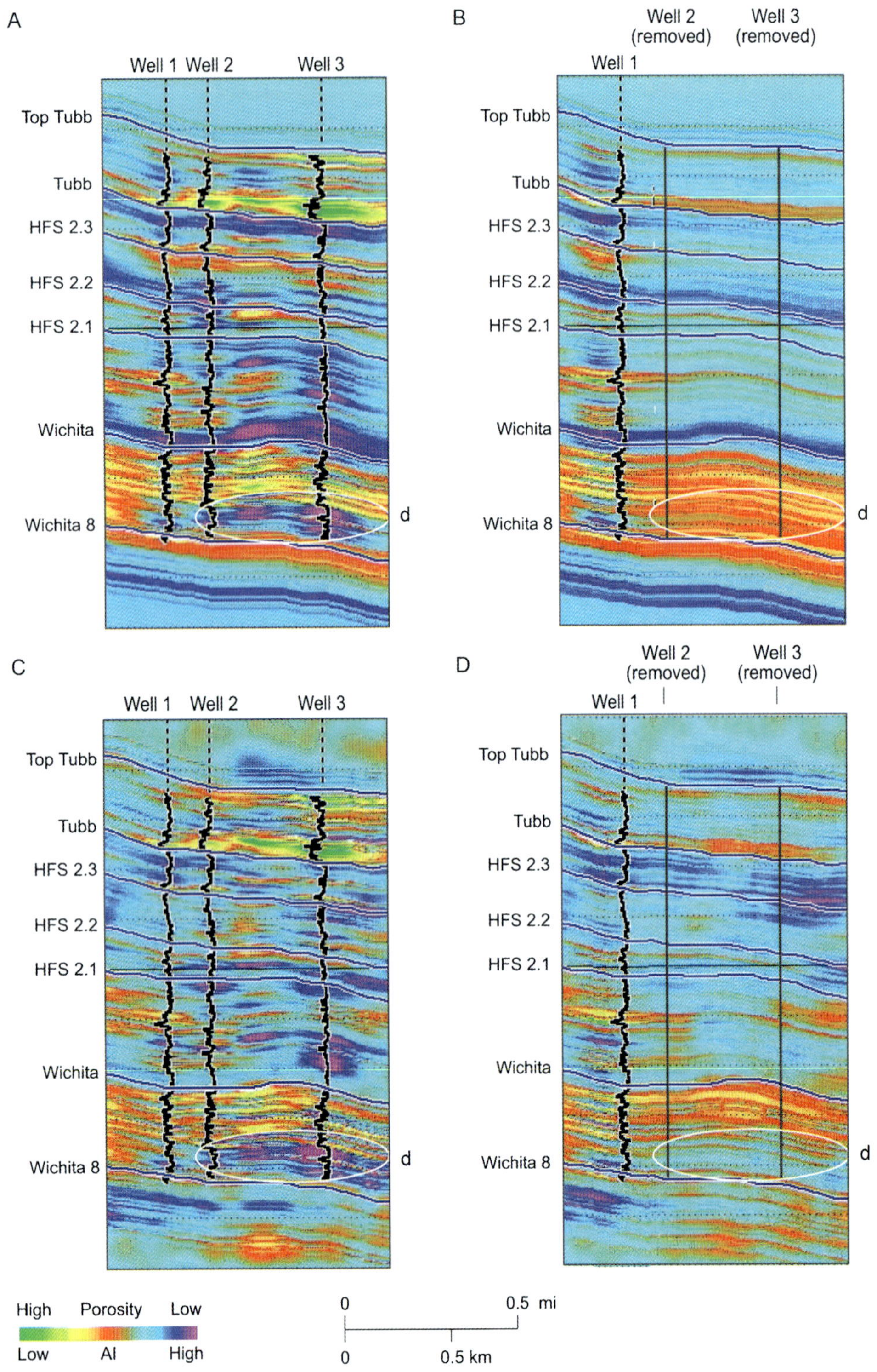

FIGURE 16. Blind test showing the value and the limit of progressive inversion (example 2). (A) Initial wireline-log acoustic impedance (AI) model. (B) Initial model with two wells removed. (C) Inversion of initial model in A with all wells. (D) Inversion of initial model in B with two missing wells. As in Figure 15, seismic inversions do a good job of imaging porosity even where wells are missing (e.g., in area d). HFS = high-frequency sequence.

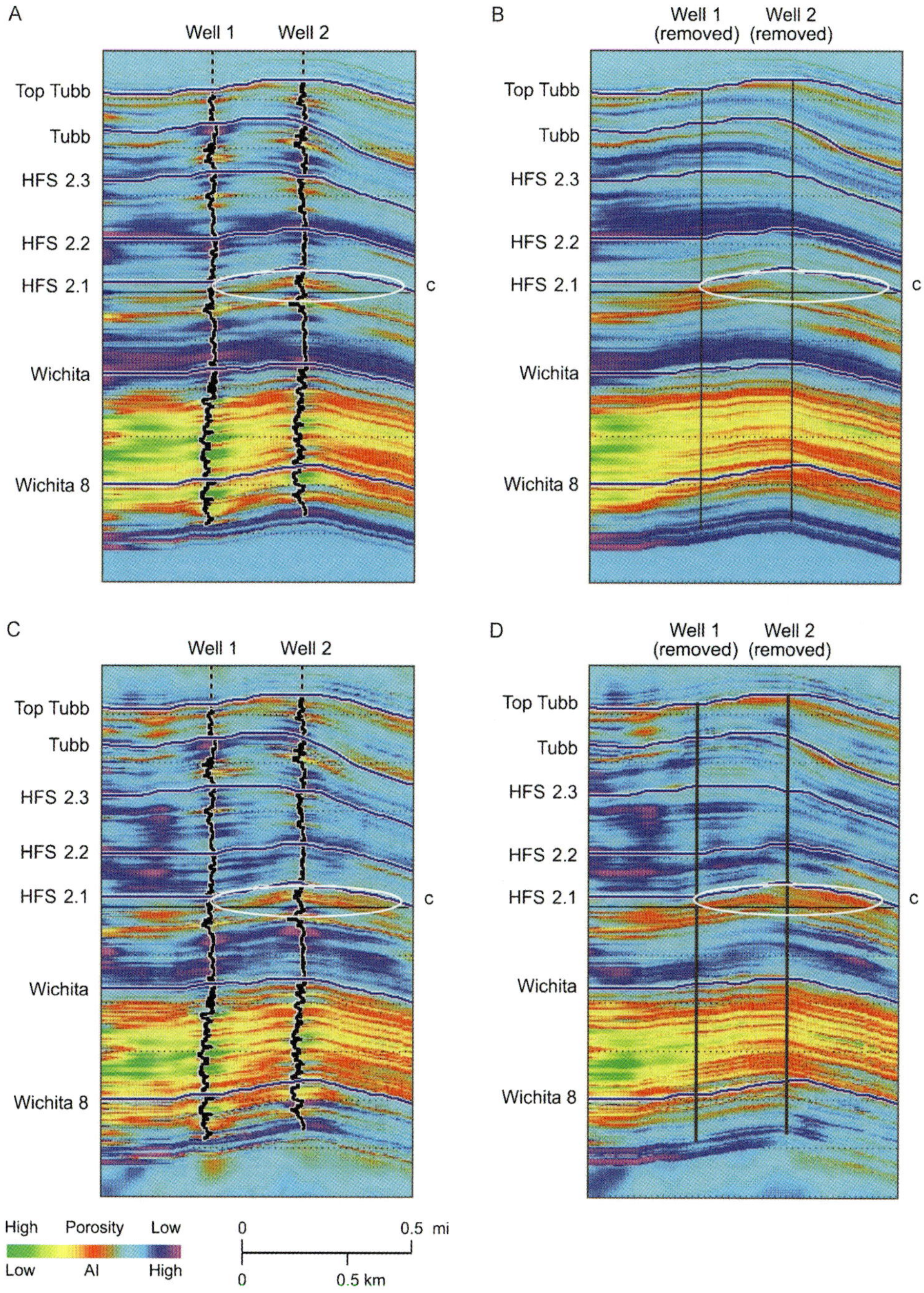

FIGURE 17. Blind test showing the value and the limit of progressive inversion (example 3). (A) Initial wireline-log based acoustic impedance (AI) model. (B) Initial model made after removing two wells. (C) Inversion of initial model in A with all wells present. (D) Inversion of initial model in B with removed wells. In this case, removing wells had little effect on model or inversion, suggesting a good control from other wells and a good agreement between well AI and seismic. HFS = high-frequency sequence.

REFERENCES CITED

Brown, A. R., 1991, Interpretation of three-dimensional seismic data, 3rd ed.: AAPG Memoir 42, 341 p.

Harrington, R. R., and F. J. Lucia, 2012, Integration of rock fabrics and stratigraphy for petrophysical quantification of reservoir framework, *in* S. C. Ruppel, ed., Anatomy of a giant carbonate reservoir: Fullerton Clear Fork (Lower Permian) field, Permian Basin, Texas: Studies in Geology 63, p. 49–66.

Ruppel, S. C., 2003, Lower Leonardian (Clear Fork-Abo) reservoir architecture: Insights from outcrops and Fullerton Clear Fork field, west Texas, *in* T. J. Hunt and P. H. Lufholm, eds., The Permian Basin: Back to basics: West Texas Geological Society Fall Symposium: West Texas Geological Society Publication 03-112, p. 45–46.

Ruppel, S. C., and R. R. Harrington, 2012, Facies and sequence stratigraphy: Critical tools for reservoir framework definition, Fullerton Clear Fork Reservoir, Texas, *in* S. C. Ruppel, ed., Anatomy of a giant carbonate reservoir: Fullerton Clear Fork (Lower Permian) field, Permian Basin, Texas: Studies in Geology 63, p. 5–48.

Tinker, S. W., D. H. Caldwell, D. M. Cox, L. C. Zahm, and L. Brinton, 2004, Integrated reservoir characterization of a carbonate-ramp reservoir, South Dagger Draw field, New Mexico: Seismic data are only part of the story, *in* G. P. Eberli, J. L. Masaferro, and J. F. Rick Sarg, eds., Seismic imaging of carbonate reservoirs and systems: AAPG Memoir 81, p. 91–105.

Tufekcic, D., J. F. Claerbout, and Z. Rasperic, 1981, Spectral balancing in the time domain: Geophysics, v. 46, no. 8, p. 1182–1188, doi:10.1190/1.1441258.

Yilmaz, O., 1987, Seismic data processing: Tulsa, Oklahoma, Society of Exploration Geophysicists, 526 p.

Zeng, H., 2003, Significance of seismic phase in interpretation of stratigraphy and sedimentology (abs.): Partners in a new environment: Canadian Society of Petroleum Geologists/Canadian Society of Exploration Geophysicists Convention, Calgary, Alberta, Canada, June 2–6, 2003, CD-ROM.

Zeng, H., M. M. Backus, K. T. Barrow, and N. Tyler, 1996, Facies mapping from three-dimensional seismic data: Potential and guidelines from a Tertiary sandstone-shale sequence model, Powderhorn field, Calhoun County, Texas: AAPG Bulletin, v. 80, no. 1, p. 16–46.

Zeng, H., and T. F. Hentz, 2004, High-frequency sequence stratigraphy from seismic sedimentology: Applied to Miocene, Vermilion Block 50 Tiger Shoal area, offshore Louisiana: AAPG Bulletin, v. 88, no. 2, p. 153–174, doi:10.1306/10060303018.

Zeng, H., S. C. Ruppel, and R. Jones, 2003, Reconditioning seismic data for improved reservoir characterization, Lower Clear Fork and Wichita, Fullerton field, west Texas, *in* T. J. Hunt and P. H. Lufholm, eds., The Permian Basin: Back to basics: West Texas Geological Society Fall Symposium: West Texas Geological Society Publication 03-112, p. 67–78.

6

Wang, F. P., and F. J. Lucia, 2012, Reservoir modeling and simulation of the Fullerton Clear Fork reservoir, Andrews County, Texas, *in* S. C. Ruppel, ed., Anatomy of a giant carbonate reservoir: Fullerton Clear Fork (Lower Permian) field, Permian Basin, Texas: Studies in Geology 63, p. 111–143.

Reservoir Modeling and Simulation of the Fullerton Clear Fork Reservoir, Andrews County, Texas

Fred P. Wang and F. Jerry Lucia

Bureau of Economic Geology, The University of Texas at Austin, Austin, Texas, U.S.A.

ABSTRACT

Simulation studies and three-dimensional (3-D) reservoir modeling were conducted as part of an integrated geologic, petrophysical, and geophysical effort to better define the distribution of remaining oil and the opportunities for a more effective recovery of remaining hydrocarbons. Two 3-D reservoir models—a 2000-ac window model and a fieldwide model—were built using a cycle-based geologic framework and rock-fabric–dependent petrophysical properties. A comprehensive sensitivity study on volumetrics was conducted using the fieldwide model, and reservoir simulation was performed in a 1600-ac area in the window model.

Original oil in place (OOIP) is a complex function of log-data quality, mapping parameters, vertical resolution of the 3-D grid, oil-water contact, and cutoff values in porosity, permeability, and water saturation. The high vertical-resolution 3-D model calculates higher OOIP than the 36-layer cycle-based model by 8 to 30%, depending on the cutoff criteria. Because permeability is a function of porosity and rock fabric, the permeability cutoff is equivalent to rock-fabric-dependent porosity or water saturation cutoffs and is less sensitive to grid vertical resolution than porosity and water saturation cutoffs.

The simulation study was divided into two phases: sensitivity analysis and history matching. The sensitivity study was used to evaluate and rank the importance of reservoir parameters affecting production performance. During simulation, oil relative permeability for primary recovery has a strong effect on recovery from waterflooding. Because fractures and breccias are common in testing and core data, negative skin factors (or effective wellbore radii) were used

DOI:10.1306/13341541St633527

to simulate near-wellbore fractures, and permeability values in the lower Wichita were modified to simulate karst-related breccias. Through history matching, optimal fluid and rock properties were determined.

INTRODUCTION

The Fullerton Clear Fork field in Andrews County, west Texas, discovered in 1942, currently produces 42° API crude from lower Wichita and Clear Fork carbonates (Table 1). The reservoir consists of tidal-flat to subtidal dolomite and limestone (Ruppel and Harrington, 2012). The field was unitized in 1953, and waterflooding operations were initiated in 1960. Production from the Fullerton Clear Fork Unit (FCU) peaked in 1986 at 15,000 BOPD and declined sharply to about 4000 BOPD in November 2008 (Figure 1). Cumulative oil production stands at about 300 million bbl.

This chapter describes the data and methodologies used in constructing 3D geologic and flow models. The key elements addressed include geologic framework, petrophysical properties, oil-water contact (OWC), volumetrics, and upscaling. The objectives of 3-D modeling were to (1) detect discrepancies among geologic, petrophysical, and engineering data; (2) build geologically realistic reservoir models; (3) establish criteria for volumetric estimations; (4) explore techniques of upscaling; (5) identify critical issues and parameters controlling production and injection fluid flow; and (6) develop strategies for reservoir management and production enhancement. Three-dimensional modeling is, therefore, an iterative process, which integrates geologic, petrophysical, geophysical, and engineering data and improves consistency among these data.

METHODS

A complete spectrum of geologic, petrophysical, geophysical, and engineering data were used to build 3-D geologic and flow models. Two reservoir models were constructed for FCU (Figure 2). A small-window model area (2000 ac) was constructed in the north-central part of the field to develop and test model construction methods and to conduct simple flow-modeling experiments. A second fieldwide model (~35,000 ac) was also constructed to define architecture, porosity, permeability, and saturation trends and to calculate reservoir volumetrics.

The small 2000-ac window model was constructed using data from 140 wells to develop an early understanding of the geology and petrophysics in an area containing the highest density of cores and best suites of wireline logs and using 20 high-frequency cycles (HFCs) from the Lower Clear Fork and Wichita formations as the geologic framework. A fine-scale geologic model containing 3.2 million cells was then constructed for the area with grid cell dimensions of 150 × 150 × 3 to 5 ft (46 × 46 × 1–1.5 m) in *x*, *y*, and *z* directions, respectively.

Although nearly 1250 wells have been completed in the unit, only those wells having good-quality wireline-log suites were used. A total of 730 wells were used in mapping and model construction. For all selected wells, petrophysical properties of porosity, permeability, and initial water saturation were calculated from core and wireline-log data (Lucia et al., 2006; Harrington and Lucia, 2012; Lucia and Kane, 2012). Two 3-D models were constructed for the fieldwide reservoir area: a coarse-grid model containing 36 layers and a high-resolution fine-grid model comprising 380 layers. Grid cell dimensions for both fine and coarse models are 150 × 150 ft (46 × 46 m) in *x* and *y* directions, respectively, resulting in about 42 million cells in the fine-grid model and 3.9 million cells in the coarse-grid model. Reservoir modeling system software was used for the construction of both models.

STRATIGRAPHIC MODELING

Core and outcrop data (Ruppel and Harrington, 2012) show that the depositional setting changed from prograding subtidal during the Abo deposition to aggradational peritidal during the Wichita deposition to aggradational to progradational subtidal during the Lower Clear Fork deposition. Ruppel and Harrington (2012) discussed the depositional and diagenetic history of the field, the controls over porosity development, and the methods used to construct the reservoir framework in detail.

Two approaches were used to develop a robust framework for the reservoir. Because the Lower Clear Fork contains abundant upward-shallowing cycles

Table 1. Geologic and reservoir parameters of the Fullerton Clear Fork Unit.*

Parameter Name	*Value*
Reservoir	Clear Fork and Wichita-Albany
Discovery date	December 1941
Unitization date	1953
Structure type	Northwest-southeast–trending anticline
Rock type	Dolomite/limestone
Primary drive mechanism	Solution gas
Secondary recovery	Waterflooding
Original oil productive area	44,000 ac
Unit area	29,541 ac
Reservoir temperature	110°F
Porosity	
Permeability	
Water salinity	90,000 ppm
API oil gravity	42° API at 60°F
Oil gravity	0.82
Oil viscosity	0.43 cp at 110°F and 2940 psi (20 MPa)
Original formation volume factor	1.6 RB/STB
Gas gravity	0.97
Original reservoir pressure	2940 psi (20 MPa) at 7200 ft (2195 m)
	2852 psi (20 MPa) at −3950 ft (−1204 m)
Original saturation pressure	2370 psi (16 MPa)
Original solution gas-oil ratio	1250 SCF/STB
1986 reservoir pressure	3100 psi (21 MPa)
1986 solution gas-oil ratio	700 SCF/STB

**Modified from Railroad Commission of Texas, 1986.*

that are relatively easy to define on wireline logs, cycle tops were used to subdivide this part of the reservoir into fifteen 10 to 15 ft (3 to 5 m)–thick layers. Cycles are less apparent in the Wichita, but diagenetically enhanced porosity development at cycle tops creates a basis for identifying and correlating many Wichita cycle tops on porosity logs (Ruppel and Harrington, 2012). Using this approach, 10 cycle-based intervals were defined for the Wichita (Figure 3).

The geologic frameworks used in the construction of the small-window and fieldwide 3-D models differ only slightly. Table 2 lists the sequences, HFCs, flow units, and numbers of subgrids in each flow unit. Because of the clinoformal nature of the underlying platform Abo deposits, a cycle-based approach was not possible for framework definition because facies vary markedly along bed surfaces and wireline logs cannot accurately define bed-top timelines. To control the distribution of model attributes in the Abo, we mimicked clinoformal geometries by arbitrarily defining downdip-stepping progradational wedges using 3-D seismic data as a guide.

Fullerton Fieldwide Model

The fieldwide 3-D model was constructed using 35 stratigraphic horizons (Table 2; Figure 3) and wireline-log data from 730 wells having good-quality data. The horizons were correlated and mapped on the basis of well picks guided by the conceptual geologic model and seismic data. Twenty-five HFC tops (Table 2; Figure 3) were identified in cores and correlated using wireline logs from the Lower Clear Fork high-frequency sequence (HFS) 2.2 to the top of the Abo (HFS 1.1-1.2

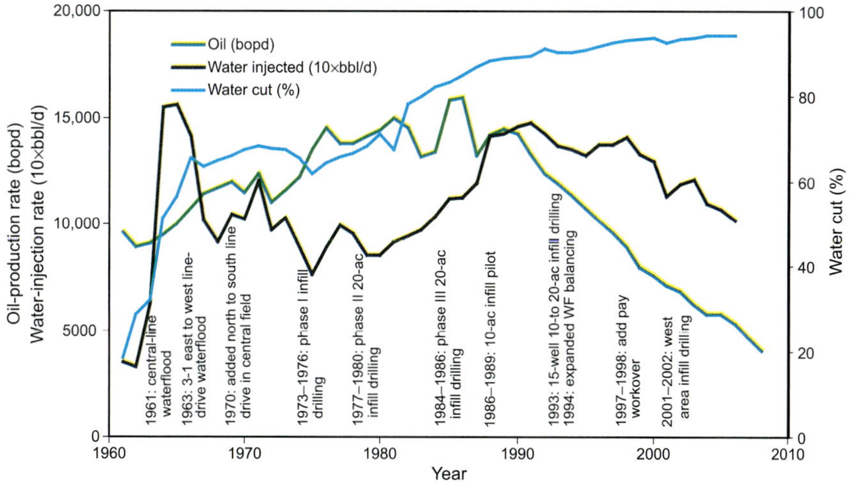

FIGURE 1. Post-1961 production history and major development events in Fullerton Clear Fork Unit.

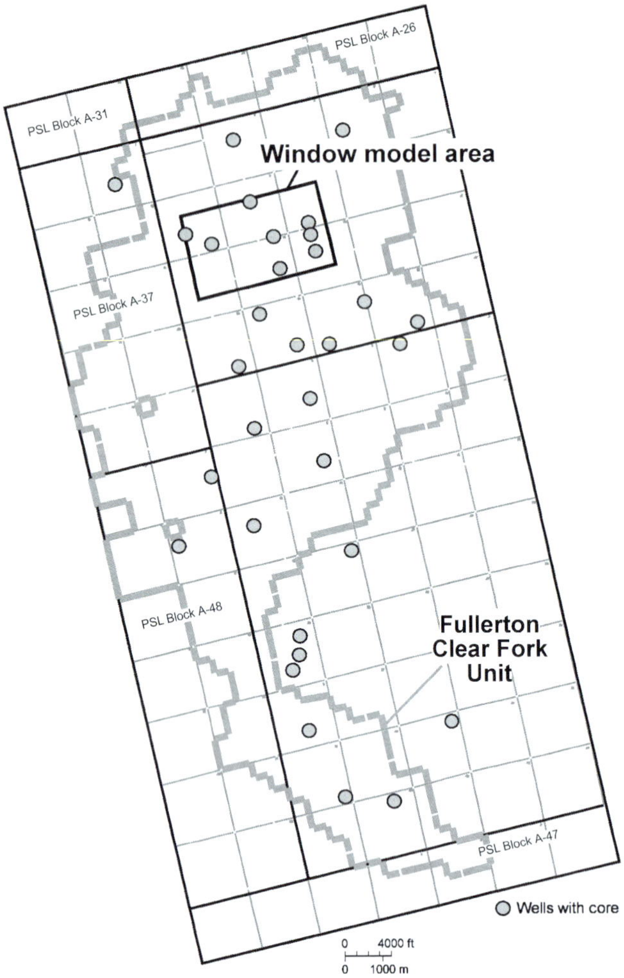

FIGURE 2. Map of Fullerton field showing the location of small-window and full-field model areas. PSL = public school land.

contact). Cycle tops were ultimately picked and correlated for a total of about 1000 wells. This included about 45,000 stratigraphic picks (R. Harrington, 2006, personal communication). However, to improve layer interpolation and limit model boundary effects, only 730 wells were actually used to distribute petrophysical data.

According to outcrop and seismic data, the Abo is a west–east-prograding outer-ramp succession (Ruppel and Harrington, 2012). However, because only scattered core data are available and limited deep wells were drilled into the Abo Formation, detailed core and log correlations necessary to model the true architecture of the Abo accurately are unavailable. Instead, 11 horizons (Table 2) were arbitrarily defined on the basis of the probable geologic architecture of the Abo (using outcrop and seismic data) and mapped and used to create a conceptual prograding-wedge model (Figure 4). The 3-D models contained grid cells that measured 150 ft (46 m) in both *x* and *y* directions. The total number of cells in the fieldwide model was about 42 million (Table 3), including 239 columns, 462 rows, and 380 layers.

Petrophysical Modeling

The rock-fabric-based approaches developed by Lucia (1983, 1999, 2007) were used to calculate permeability and water saturation from wireline-log data at individual wells. Permeability in carbonates is a function of rock fabric, pore type, and pattern of cementation (Lucia, 1983; Wang et al., 1998; Harrington and Lucia, 2012; Lucia and Kane, 2012). In many unsubdivided carbonate successions, permeability can range from two to three orders of magnitude for a given porosity. Recognition of rock-fabric classes allows carbonates to be subdivided into

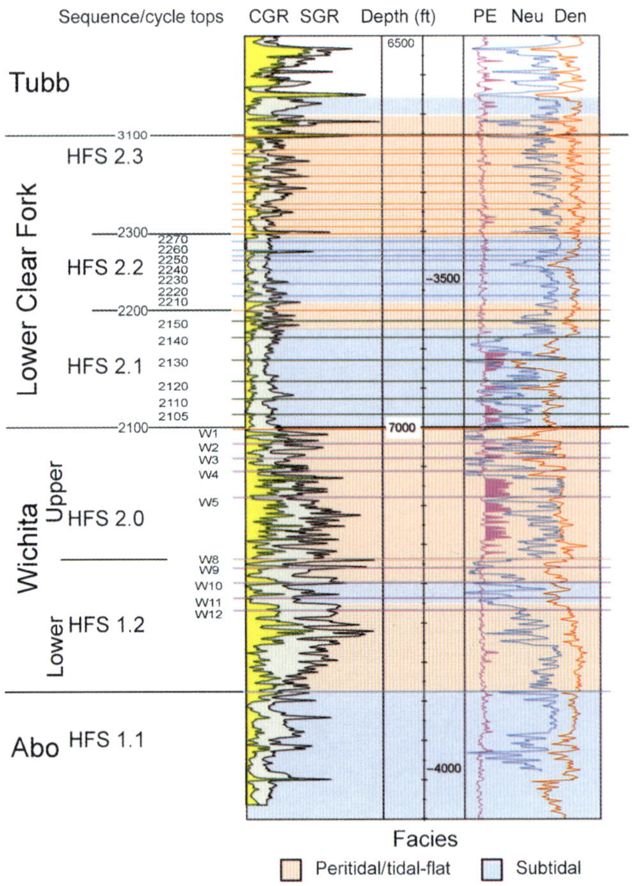

FIGURE 3. Type log of producing reservoir section at Fullerton Clear Fork field showing the relationships between formation names, depositional sequences, and reservoir zones. HFS = high-frequency sequence; CGR = computed gamma ray log; SGR = spectral gamma ray log; PE = photoelectric log; Neu = neutral log; Den = density log.

genetic groups that have more consistent porosity and permeability relationships, thus permitting better permeability estimates (Harrington and Lucia, 2012).

Strictly speaking, rock-fabric-based porosity and permeability relationships are most valid in carbonate successions having interparticle pores and dispersed cements. Successions having vuggy pores and more extensive cements are more complicated and require corrections to the rock-fabric model. The presence of separate-vug pores, for example, results in an increase in porosity, with little to no increase in permeability (Lucia and Conti, 1987; Wang et al., 1998; Harrington and Lucia, 2012). Fractures and touching vugs, however, increase permeability significantly, with only a slight increase in porosity. The pattern of cementation can also affect porosity and permeability correlations (Harrington and Lucia, 2012). The presence of poikilotopic anhydrite typically reduces porosity more than permeability. Although counterintuitive, the presence of poikilotopic anhydrite in a grain-dominated dolostone results in a better porosity-permeability relationship than would be expected from the depositional facies texture (Harrington and Lucia, 2012).

Petrophysical properties of rock fabric, porosity, permeability, and water saturation from 730 selected wells were loaded into the RMS modeling software. Petrophysical properties were first upscaled to a grid resolution of 3-D models and mapped using the inverse distance technique with a search radii of 2500 (762 m) and 5000 ft (1524 m). Figure 5 shows porosity distributions in two layers, with a search radii of 2500 (762 m) and 5000 ft (1524 m). Because the western and southern parts of the unit (circled areas in Figure 5) contain older wells of relatively poor log quality, the spacing of usable wells in this area is sparse. Because of the limited control in these areas, the search radius had to be widened to 5000 ft (1524 m) to increase mapping accuracy.

Porosity Distribution

Porosity distribution in the Fullerton Clear Fork reservoir is a complex function of facies, rock fabric, lithology, and diagenesis overprints (Ruppel and Harrington, 2012). The 3-D model (Figure 6) shows that the porosity in the field is highly heterogeneous, both laterally and vertically.

The upper Lower Clear Fork HFS 2.2 is a highly cemented grain-rich subtidal interval with less than 10% porosity values. Porosity is better developed in the southern end of the field. Only a northwest-southeast–trending moderate porosity trend is developed in the northern part of the field (Figure 7) in cycles L2220 to L2260.

The Lower Clear Fork HFS L2.1 is a grain-dominated subtidal unit much like HFS 2.2, although the average porosity in this interval is mostly higher (Figure 7). In the northern part of the field, porosity is best developed in a northwest–southeast trend, mostly in cycles L2105 to L2140 (Figure 8). Reservoir quality decreases away from this high-porosity trend sharply, and porosity in most of the western part of the field is low. A laterally extensive, low-permeability succession of tidal-flat cycles near the contact of sequences L2.1 and L2.2 (cycles L2220 and L2150; Figure 8) separates higher porosity intervals in the two sequences. No such continuous flow barriers exist between HFS L2.1 (lower Lower Clear Fork) and the underlying uppermost Wichita (HFS L2.0).

The Wichita includes two HFSs: the upper Wichita (L2.0) and the lower Wichita (L1.2). Both of these sequences are dominated by mud-rich, inner-ramp tidal-flat facies. Porosity in the upper Wichita (L2.0) is significantly higher than in the grain-rich subtidal rocks of HFS 2.2 (upper Lower Clear Fork) and slightly higher than in the similarly grain-rich rocks of HFS 2.1 (lower Lower Clear Fork). Porosity in upper Wichita cycles (L2.0, W1 and W2) displays a distribution that is similar to that in HFS 2.1 (Figure 8). It is best developed in the northeastern part of the field and decreases sharply to the south and west. Wichita porosity is much more widely scattered in the lower upper Wichita (L2.0, W4 and W5) cycles. Porosity is generally high and evenly distributed in the intervals of the lower Wichita (L1.2, W8–W11). The low-porosity lower L2.0 succession (cycles W4 and W5) in the northern part of the field serves as a barrier to separate the Wichita into upper (L2.0, W1–W3) and lower (L1.2, W8–W12) porosity intervals. In the southern part of the field, no such mid-Wichita low-porosity zone is present. Overall, reservoir porosity is best developed in the Lower Clear Fork (HFS 2.1), in the upper part of Wichita HFS 2.0, and in the upper part of Wichita HFS 1.2 (Figure 8).

Permeability and Horizontal Permeability Distributions

Reservoir data can be considered to comprise five scales for purposes of reservoir modeling: wireline

Table 2. Geologic frameworks used in three-dimensional reservoir models.

Formation		*High-Frequency Sequences*	*Reservoir Unit*	*Small-Window Model*			*Fieldwide Model*	
				Horizon	*Flow Unit*	*Layer Number*	*Cycle*	*Layer Number*
Clear Fork	Lower	HFS* 2.2	Upper	LC1	LC1	1	2300 (not included)	
							2270	1
					LC1a	2	2260	2
				LC2	LC2	3	2250	3
							2240	4
					LC2a	4	2230	5
				LC3	LC3	5	2220	6
					LC3a	6	2210	7
				LC4	LC4	7	2200	8
					LC4a	8		
		HFS* 2.1	Lower	LC5	LC5	9	2150	9
					LC5a	10		
				LC6	LC6	11	2140	10
					LC6a	12		
				LC7	LC7	13	2130	11
					LC7a	14		
				LC8	LC8	15	2120	12
					LC8a	16		
				LC9	LC9	17	2110	13
					LC9a	18		
				LC10	LC10	19	2105	14
Wichita	Upper	HFS* 2.0	Upper	W1	W1	20	W1	15
					W1a	21		
				W2	W2	22	W2	16
					W2a	23		
				W3	W3	24	W3	17
					W3a	25		
			Lower	W4	W4	26	W4	18
					W4a	27		
				W5	W5	28	W5	19
					W5a	29		
	Lower	HFS* 1.2	Upper	W8	W8	30	W8	20
					W8a	31		
				W9	W9	32	W9	21
					W9a	33		
				W10	W10	34	W10	22
					W10a	35		
				W11	W11	36	W11	23
					W11a	37		
			Lower	W12	W12	38	W12	24
					W12a	39		
Abo		HFS* 1.1		Abo	Not included	Not included	Abo	25
							Abo10	26
							Abo9	27
							Abo8	28

Table 2. cont.

Formation	*High-Frequency Sequences*	*Reservoir Unit*	*Small-Window Model*			*Fieldwide Model*	
			Horizon	*Flow Unit*	*Layer Number*	*Cycle*	*Layer Number*
						Abo7	29
						Abo6	30
						Abo5	31
						Abo4	32
						Abo3	33
						Abo2	34
						Abo1	35
						Abo1 + 100 ft	36

**HFS = high-frequency sequence.*

logs, high-resolution layers, flow units, HFCs, and the entire reservoir. Rock-fabric-based flow units are the building blocks used in a cycle-based geologic framework. High-frequency cycles in the Fullerton reservoir typically have high-porosity–high-permeability, grain-dominated fabrics at cycle tops, and low-porosity–low-permeability, mud-dominated fabrics at cycle bases. However, the relationship between permeability distribution and facies-stacking patterns differs significantly at different reservoir scales. These differences must be considered in vertical upscaling for accurate model construction.

Upscaling

Permeability was calculated from porosity and rock-fabric-based data at wireline-log resolution for each well. Wireline-log-scale permeability was then upscaled vertically to model scale resolution using an arithmetic averaging technique before being mapped into a 3-D space. A detailed discussion of vertical upscaling techniques, including the arithmetic averaging versus the geometric averaging and the crossflow effect, is given in Appendix I.

Lucia (1983, 2007) showed that, for a given porosity, permeability can be an order of magnitude higher for a carbonate grainstone (class I rock fabric) than that for a mud-dominated rock (class III rock fabric). As this relationship predicts and as Harrington and Lucia (2012) show, permeability values are significantly lower in the mud-dominated peritidal facies of the Wichita than in the subtidal facies of the Lower Clear Fork despite the generally high porosity common to the former.

Figures 9 and 10 show the permeability distribution, and Figure 11 shows the horizontal permeability (k_h) distribution in the field. Because permeability and k_h are better displayed on a logarithmic scale,

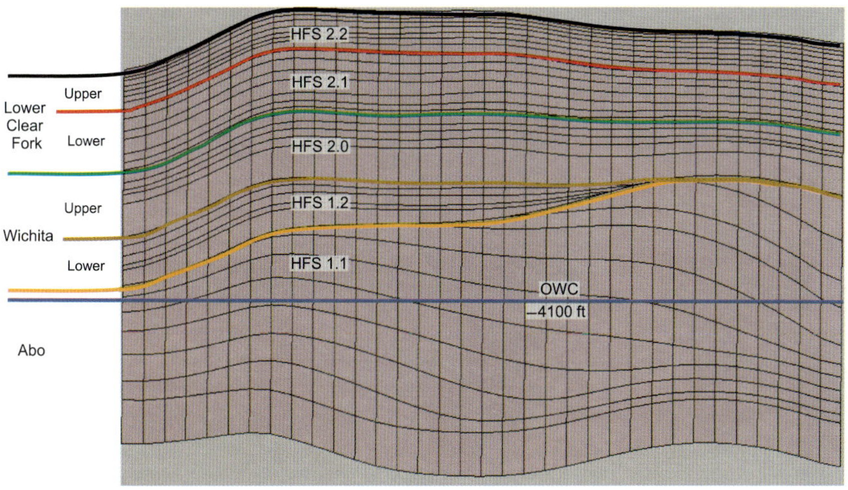

FIGURE 4. Cycle-based geologic framework used to build three-dimensional reservoir models at Fullerton Clear Fork field. The models include four major horizons and 35 high-frequency cycles. Abo was modeled with 11 east-dipping clinoformal layers. HFS = high-frequency sequence; OWC = oil-water contact.

Table 3. Dimensions of three-dimensional field and small-window models.

Model	*Field*		*Small Window*	
Item Name	*Dimension (ft)*	*Number of Cells*	*Dimension (ft)*	*Number of Cells*
X	150	239	150	150
Y	150	462	150	90
Z	Varying	380	3–5	256
Total		42×10^6		3.2×10^6

special procedures were followed to generate average-permeability and k_h maps on the logarithmic scale. Permeability and k_h in the fine-grid model were upscaled to a seven-layer zone model using the arithmetic mean. Permeability and k_h maps were generated from logarithmic values of upscaled permeability and k_h. Average permeability and k_h in the Fullerton reservoir are significantly higher in the Lower Clear Fork (HFS 2.1 and L.2.2) than in the Wichita (HFS 2.0 and L.1.2) (Figures 10, 11). With the highest permeability and k_h, Lower Clear Fork HFS 2.1 has shown the highest waterflooding recovery efficiency.

Anomalous water production has been reported from the Wichita by the field operator. Although not predicted or explained by matrix porosity or permeability data, these observations may be caused by the presence of zones of karst breccias in the Wichita. Breccias and related fractures have been reported from the middle Wichita in several cores in the field (Ruppel and Harrington, 2012). Alternatively, anomalous water production may be associated with fractures induced by high-pressure water injection, a common problem in high-porosity–low-permeability reservoir zones.

Water Saturation and Soϕh Distribution

The techniques used to calculate water saturation in the Fullerton reservoir are discussed by Lucia and Kane (2012). One of the major petrophysical questions in the Fullerton reservoir is the validity of resistivity-log measurements for calculating saturation. In many instances, resistivity-log calculations of water saturation based on Archie equation calculations are lower than water saturation values derived from capillary-pressure data (Lucia and Kane, 2012). This is especially apparent in, for example, low-porosity grain-dominated rocks of the Lower Clear Fork HFS 2.2 and the Abo. Oil saturations derived from the resistivity-log data are very high in these zones. As pointed out by Lucia and Kane (2012), these observations suggest that water saturation estimates using the Archie equation are not reliable when porosity is less than 5%. For this reason alone, water saturation calculated from capillary-pressure modeling were used in the models at Fullerton.

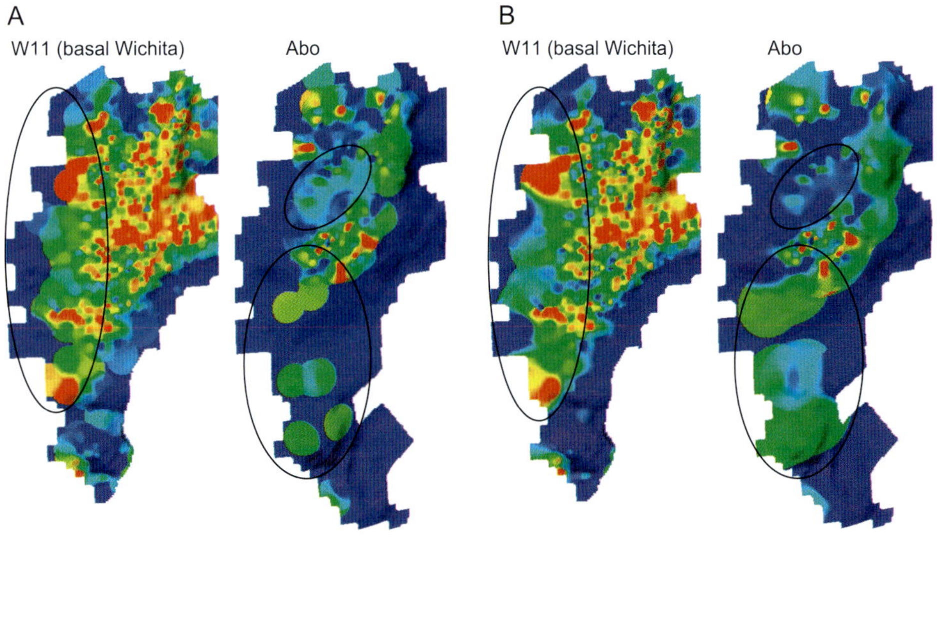

FIGURE 5. Effect of search radius on porosity mapping. (A) 2500 ft (762 m). (B) 5000 ft (1524 m). The circled areas show that a search radius of 5000 ft (1524 m) improves mapping coverage. Models show, however, that large search radii can lead to an overestimation of reservoir volumetrics.

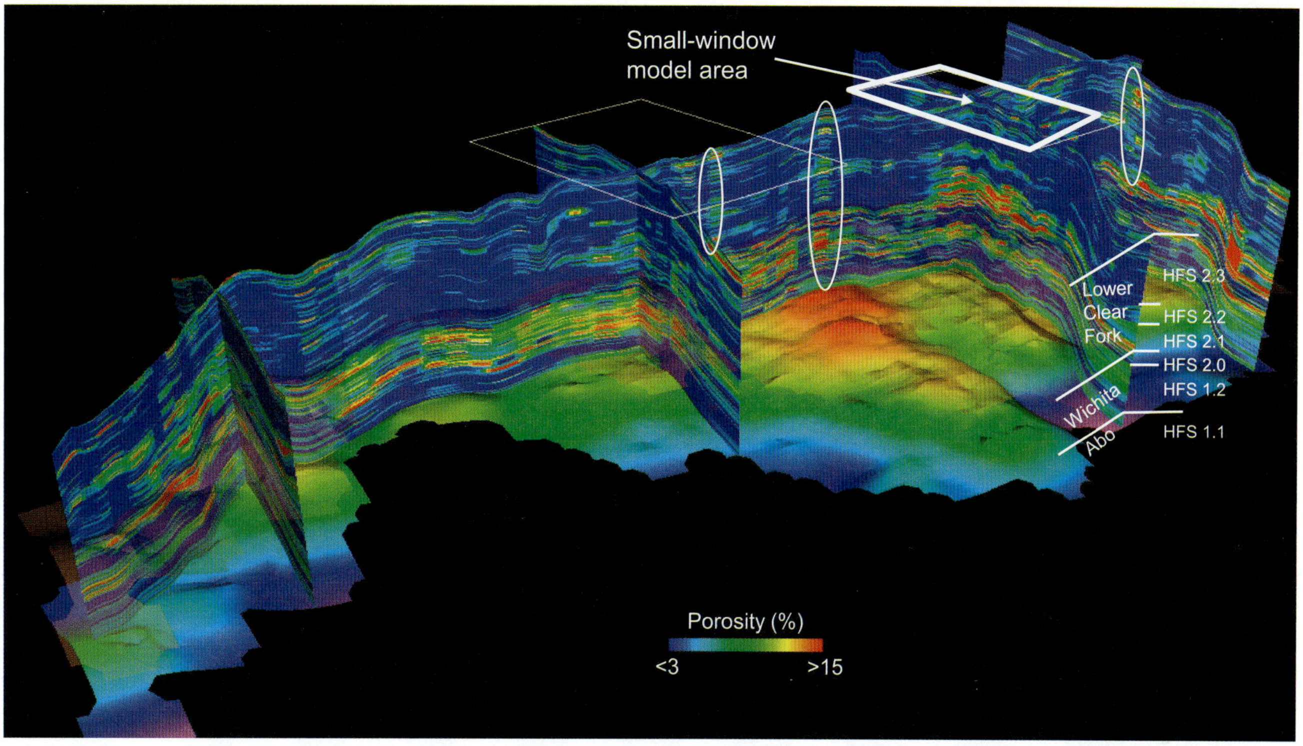

FIGURE 6. Distribution of porosity in full-field three-dimensional model showing the study areas and the quality of wireline-log data wells. The rectangle was the small-window model area for flow simulation, and the ovals were the wells with a very high porosity. Note that high porosity is very common in the Wichita over large areas of the field. HFS = high-frequency sequence.

Figure 12 depicts the distribution of initial oil saturation in the fieldwide 3-D model. Distributions of oil saturation and $S_o\phi_h$ by sequence are shown in Figures 13, 14. The Lower Clear Fork (HFS 2.1 and HFS 2.2) contained 57% of original oil in place (OOIP), whereas the Wichita contained 43% (Figure 15). With 18% of OOIP calculated in the Lower Clear Fork HFS 2.2, most oil is in the southern part of the field. The highest hydrocarbon pore volumes in the unit are found in cycles L2140, L2130, L2120, L2110, L2105, W1, and W5 because of their high porosity values (Figures 3, 8).

VOLUMETRICS

One of the purposes of 3-D geologic models is to provide an accurate and flexible way of estimating OOIP, which is calculated from 3-D models by:

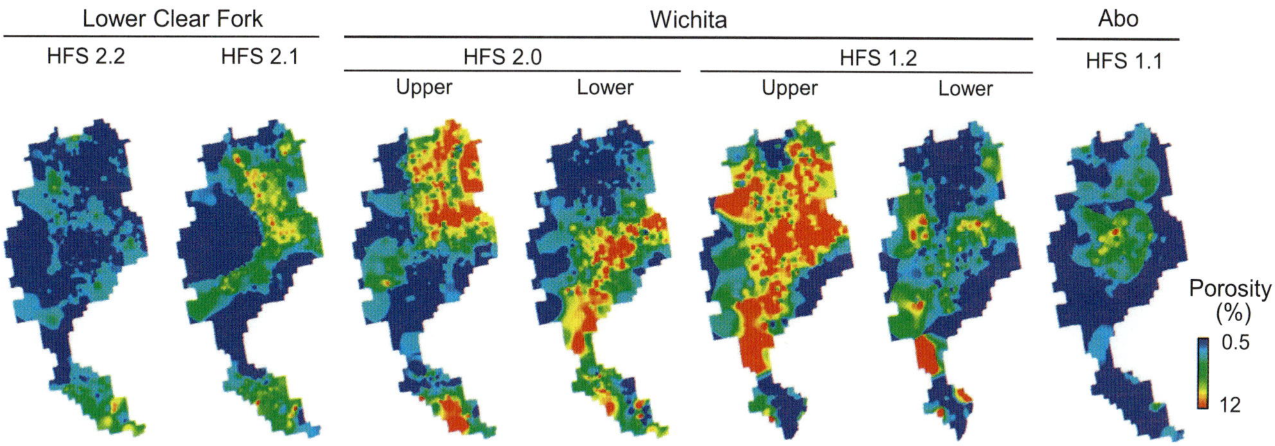

FIGURE 7. Distribution of average porosity in high-frequency sequences (HFSs) of the Lower Clear Fork, Wichita, and Abo at Fullerton field.

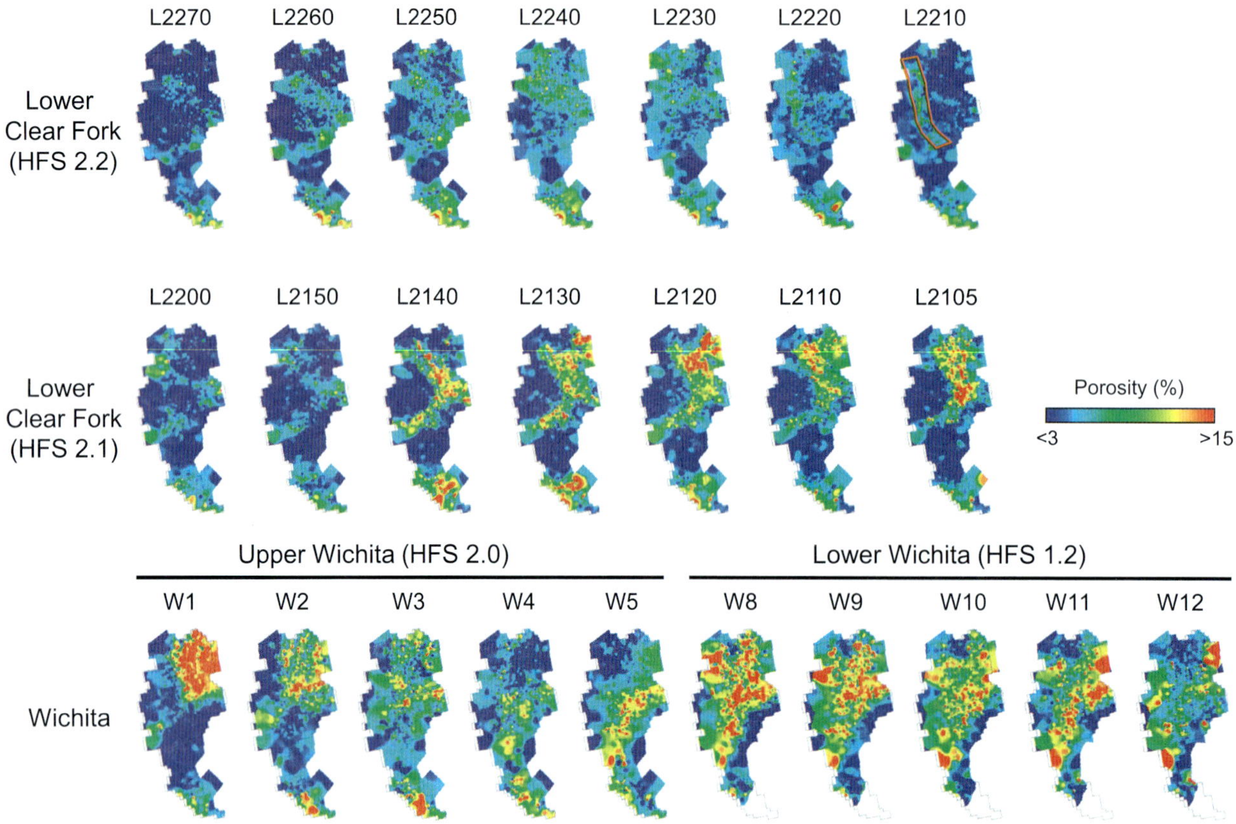

FIGURE 8. Distribution of average porosity in high-frequency cycles in the Lower Clear Fork and Wichita. HFS = high-frequency sequence.

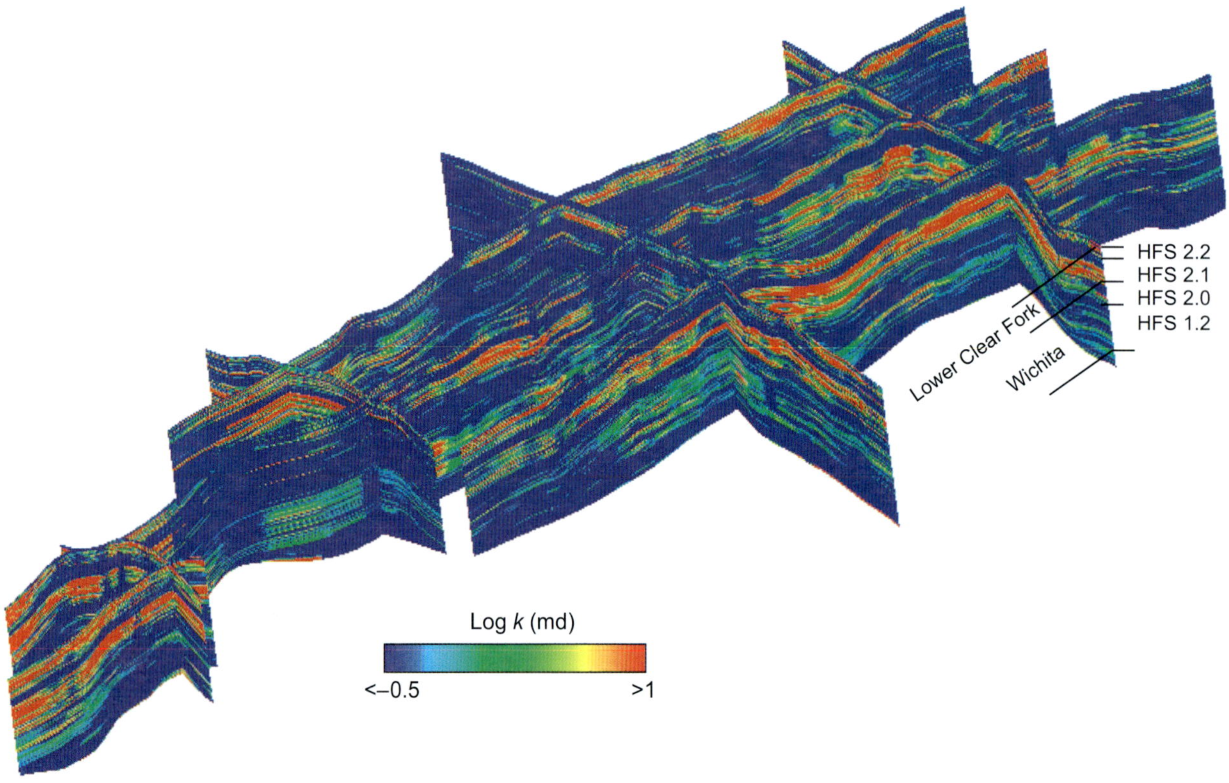

FIGURE 9. Distribution of permeability in full-field three-dimensional model. Note that high permeability is most common in the Lower Clear Fork (cf. with porosity distribution in Figure 4). HFS = high-frequency sequence.

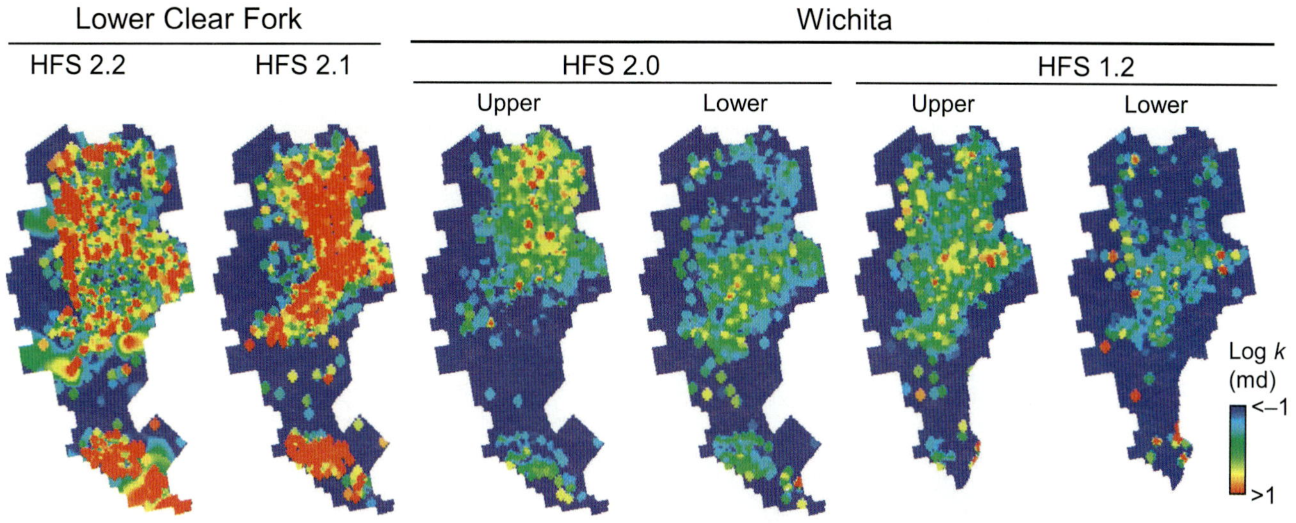

FIGURE 10. Distribution of average permeability in high-frequency sequences (HFSs) in the Lower Clear Fork and Wichita.

$$V_o = 0.1781B_o \sum_{k=1}^{kk} \sum_{j=1}^{jj} \sum_{i=1}^{ii} A_{ijk} h_{ijk} \varphi_{ijk} (1 - S_{w,ijk}) pay_{ijk}, \quad (1)$$

where V_o = OOIP (BSTB); B_o = formation volume factor (bbl/STB); A_{ijk} = area (ft^2); h_{ijk} = thickness (ft); φ_{ijk} = porosity (fraction); $S_{w,ijk}$ = water saturation (fraction); pay_{ijk} = pay flag, 0 or 1.

In equation 1, the pay flag is determined by cutoff values. One is assigned to the pay flag if the property value is equal to or greater than the cutoff value and vice versa. The previous study by Exxon Southwest/Rocky Mountain Division (1985) reported two types of oil in the Fullerton Clear Fork reservoir with formation volume factors of 1.615 and 1.36 (Figure 16). However, because the areal and vertical coverage of the second oil type is not clearly determined and is thought to be relatively small, a formation volume factor of 1.6 (Table 4) was used in this study for the entire reservoir.

Major sources of error in estimations of reservoir volumetrics are wireline-log data (porosity and water saturation), limitations of petrophysical models, OWC, mapping parameters, and vertical resolution of 3-D models. For mature hydrocarbon fields such as the Fullerton Clear Fork field, many wells were drilled before the arrival of modern porosity and resistivity-logging tools. Accordingly, accurate modern porosity logs are of limited availability, and OOIP estimates can thus vary significantly. A sensitivity analysis was performed to determine the most likely range of OOIP values for the FCU. The parameters studied include search radius in mapping, grid vertical resolution,

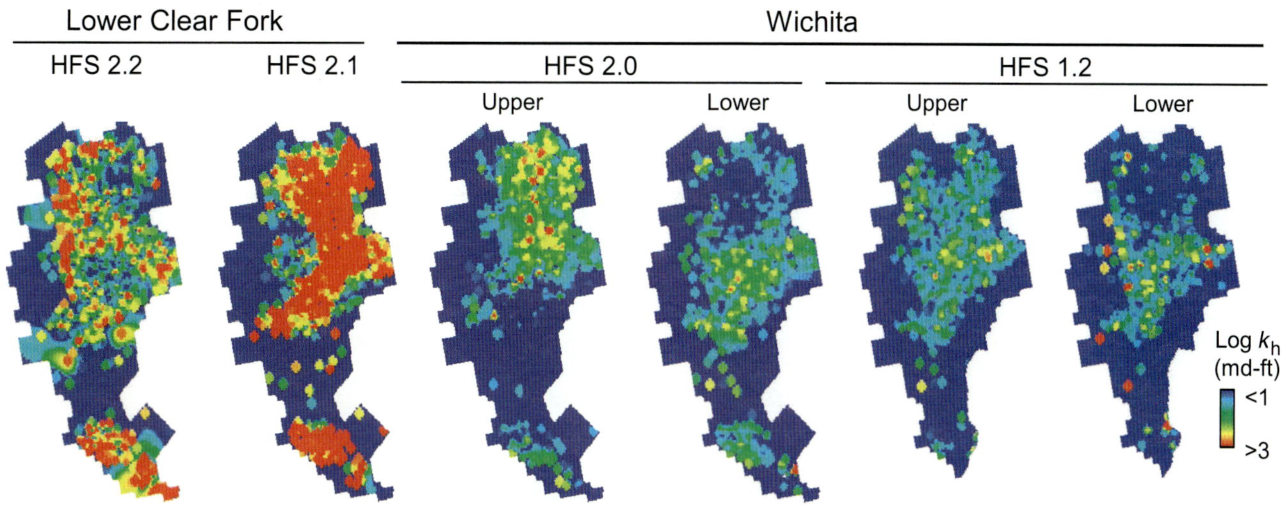

FIGURE 11. Distribution of k_h in high-frequency sequences (HFSs) in the Lower Clear Fork and Wichita.

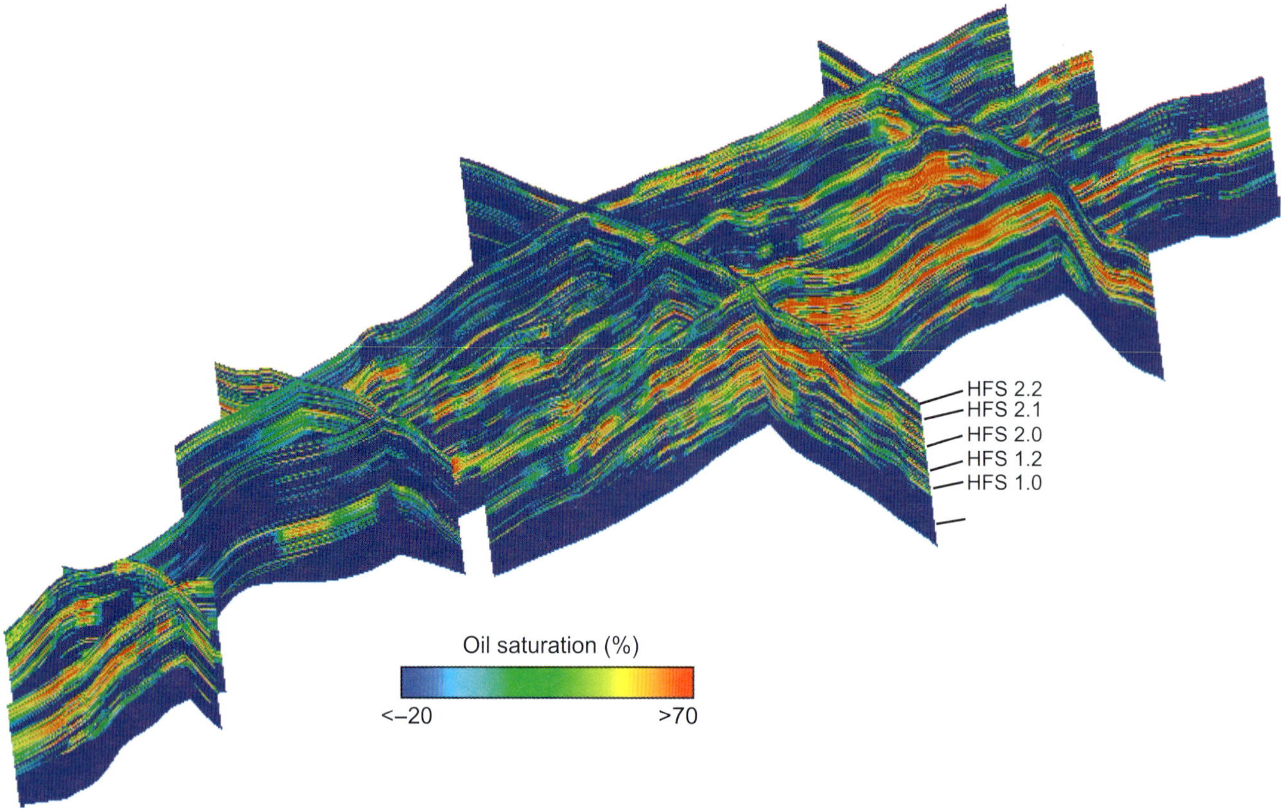

FIGURE 12. Distribution of oil saturation in full-field three-dimensional model. Note that high oil saturations are most common in the Lower Clear Fork (based on oil-water contact at −3940 ft [−1201 m]). HFS = high-frequency sequence.

OWC, and cutoff criteria for porosity, water saturation, and permeability.

Search Radius in Mapping

As discussed previously, search radius affects OOIP estimates by controlling the smoothness and coverage of petrophysical properties. In areas with close well spacing, a small search radius will suffice for accurate mapping. By contrast, in areas or intervals with wide well spacing and, thus, limited well control, large search radii can create maps that can either underestimate or overestimate OOIP. The effect of change in mapping radii is shown graphically in Figure 5. In both the Wichita and the Abo, large radii mapping provides better map coverage of area of poor well control. However, in both cases, this resulted in some

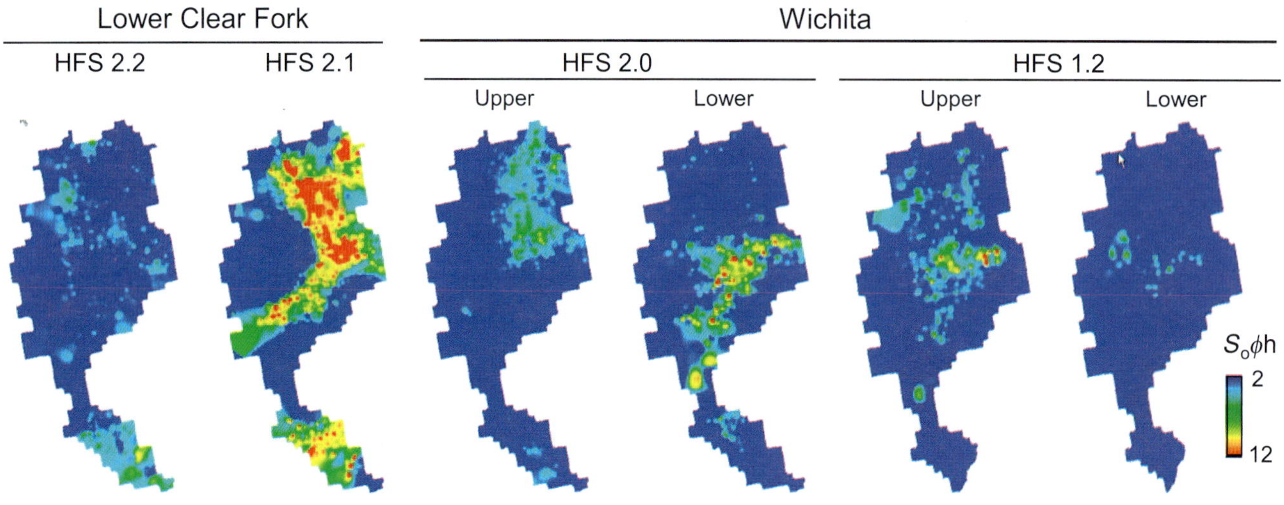

FIGURE 13. Distribution of oil saturation in high-frequency sequences (HFSs) in the Lower Clear Fork and Wichita.

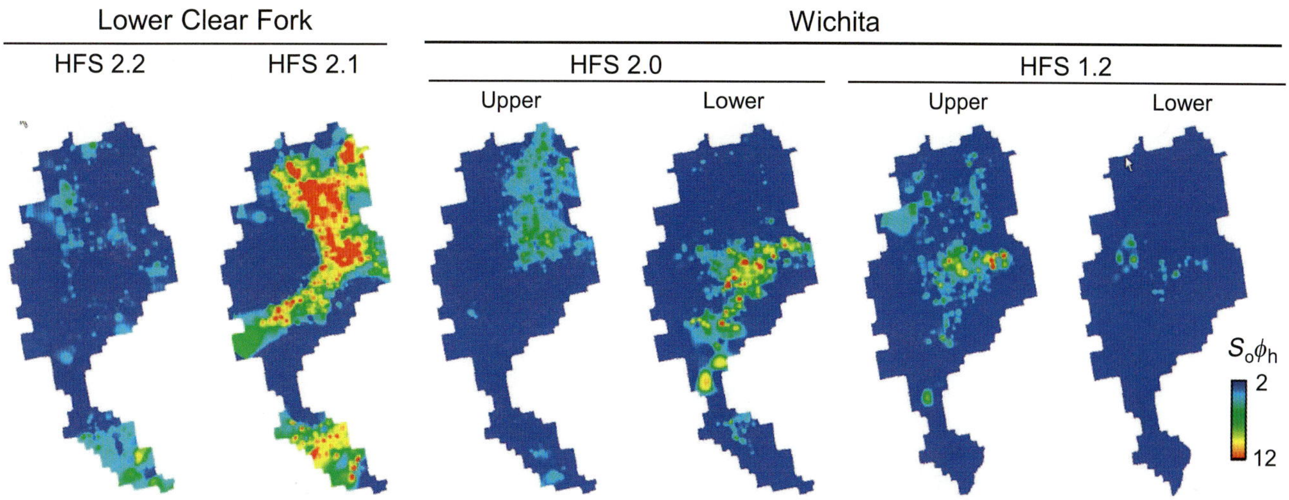

FIGURE 14. Hydrocarbon volume ($S_o\phi_h$) distributions in high-frequency sequence (HFS) 2.2, HFS 2.1, HFS 2.0, HFS 1.2, and HFS 1.1.

increases in the mapped porosity. As shown in Table 5, increasing the search radius from 2500 (762 m) to 5000 ft (1524 m) resulted in a 7% increase in OOIP calculations for the entire field and 4 to 20% increases in OOIP calculations for the individual unit. The largest 20% increase in OOIP is observed in the Abo, where well control is generally poor. With uncertainties high in the Abo, an intermediate search radius of 2500 ft (762 m) could reduce the risk of becoming overly optimistic.

Grid Resolution

Vertical resolution of the 3-D model also has a significant impact on OOIP estimates. Optimal vertical resolution of the 3-D model is the coarsest resolution that has a minimal effect on the OOIP estimate. I tested the effect of vertical resolution on 3-D models of OOIP using a sequence-based model, a cycle-based model, and a 380-layer fine-grid model. The calculated OOIP values were 1.582, 1.405, and 1.21 BSTB for the fine-grid, the flow-unit-based, and the coarse-grid cycle-based models, respectively. The 23% difference in OOIP between coarse- and fine-grid models suggests that fine-grid models should be preferred for OOIP calculations.

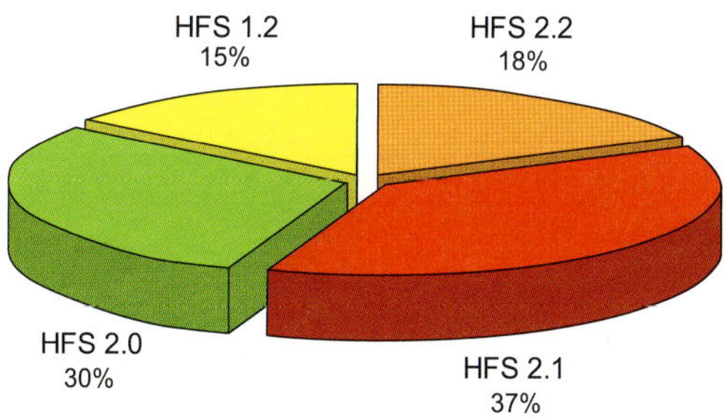

FIGURE 15. Relative distribution of hydrocarbon volume ($S_o\phi_h$) in high-frequency sequences (HFSs).

Oil-Water Contact

As discussed previously, low porosities in the Abo at the bottom of the Fullerton reservoir result in unreliable log-resistivity values, making it difficult to determine the true OWC from wireline-log data. Accordingly, four OWC contacts (−3940 ft [−1201 m], −4000 ft [−1219 m], −4050 ft [−1234 m], and −4100 ft [1250 m]) were tested to study the effect of OWC on OOIP. The −3940 ft (−1201 m) contact is the shallowest OWC reported in previous studies by the operator. An OWC of −4100 ft (−1250 m) was used in this study to estimate original water saturation from capillary-pressure data. This value is supported by core saturation data in a downdip core in the field (Amoco FM-1). The OOIP values (total pore volume) range from 1.582 BSTB for the OWC case at −3940 ft (−1201 m) to 1.917 BSTB for the OWC case at −4100 ft (−1250 m) (Table 6; Figure 17).

In the case of the OWC at −4100 ft (−1250 m), a significant part of the Abo is included in the hydrocarbon volume calculations. In the case of OWC at −3940 ft (−1201 m), almost the entire Abo reservoir is eliminated, and the hydrocarbon volume in the Wichita is reduced by about 13%. Although the estimated OOIP in the Abo is 236 MMSTB for the case of OWC at −4100 ft (−1250 m), uncertainties in porosity and water saturation and OOIP estimates are high because of the lack of well control.

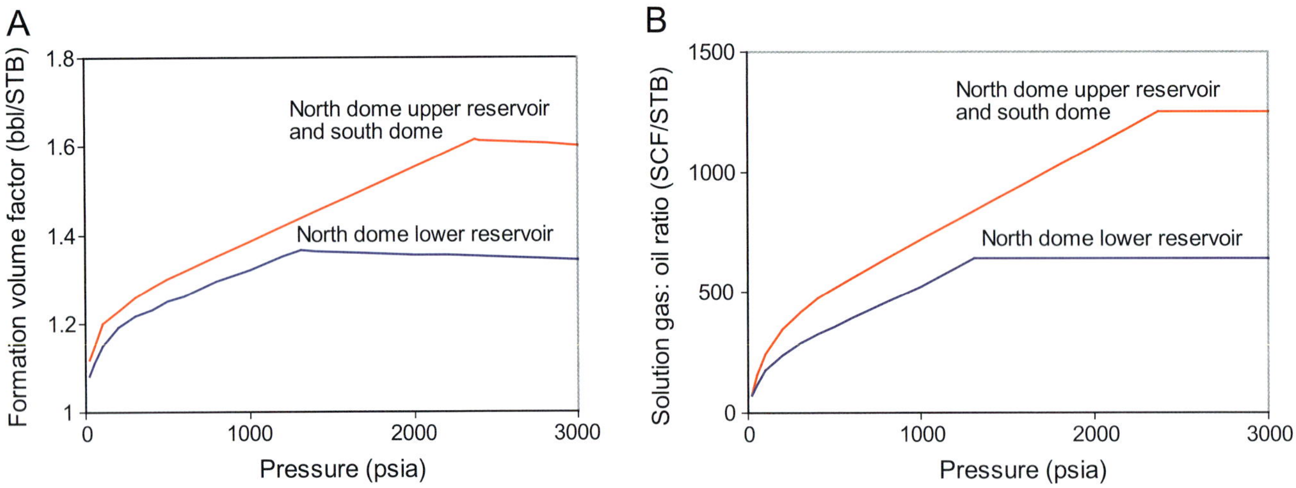

FIGURE 16. (A) Formation volume factor and (B) oil-gas ratio of two types of crude in Fullerton Clear Fork Unit. The red curves are for the crude oil from the upper reservoir including high-frequency sequence (HFS) 2.0, HFS 2.1, and HFS 2.2, and the black curves are for the crude oil from the lower reservoir including HFS 1.1 and HFS 1.2.

Cutoff Criteria

Cutoff values are used to remove non-reservoir-quality rocks to reduce uncertainties in petrophysical equations at low-porosity intervals and to eliminate high water-bearing intervals from averaging petrophysical properties. Sensitivity studies were conducted on porosity, permeability, and saturation to evaluate the impact on OOIP.

Porosity Cutoffs

Four porosity cutoff values were considered: 0.03, 0.05, 0.07, and 0.10 (Table 7; Figures 18, 19). Note that the porosity cutoff value should vary with rock fabric. As Lucia (2007) has shown (Harrington and Lucia, 2012; Lucia and Kane, 2012), permeability and water saturation are lower in grain-dominated fabrics than in mud-dominated fabrics at any given porosity value. Therefore, porosity cutoff values used for the Lower Clear Fork should be lower than those for the Wichita. Figure 18A to C shows the effect of porosity cutoff values 0.03, 0.05, and 0.1 on porosity distributions. Figure 19A shows that the OOIP value decreases gradually with increasing porosity cutoff values from 0 to 0.05, and the rate of decrease increases sharply when the porosity cutoff value exceeds 0.05.

Water Saturation Cutoffs

The purpose of water saturation cutoff values is to remove errors and uncertainties in porosity, resistivity, and water saturation models in low-porosity intervals and to remove nonproductive high water saturation transition-zone intervals. As shown in Figure 19B, OOIP values increase with higher water saturation cutoffs from 1184 BSTB when the cutoff value of water saturation is 50% to 1.557 BSTB when the cutoff value

Table 4. Formation volume factor, oil-gas ratio, and oil-water contact used in original-oil-in-place estimates.

Parameter	*Value*	
Formation volume factor (bbl/BBL)	1.6	
Gas-oil ratio (SCF/STB)	1250	
	Operator	This Study
Oil-water contact (ft)	−3940	−4100

Table 5. Effect of search radius on original-oil-in-place estimates of Fullerton Clear Fork from fine-grid model with oil-water contact at −4100 ft.

Search Radius	*2500 ft*	*5000 ft*
	Original Oil in Place (BSTB)	
Lower Clear Fork HFS* 2.2	0.2591	0.2791
Lower Clear Fork HFS* 2.1	0.5744	0.5995
Upper Wichita HFS* 2.0	0.2114	0.2173
Lower Wichita HFS* 2.0	0.2672	0.2779
Upper Wichita HFS* 1.2	0.2038	0.2164
Lower Wichita HFS* 1.2	0.0777	0.0907
Subtotal	1.5936	1.6810
Abo HFS* 1.1	0.1969	0.2357
Total	1.7905	1.9167

**HFS = High-frequency sequence.*

Table 6. Effect of oil-water contact on original-oil-in-place estimates of the model area from fine-grid model.

Oil-Water Contact	*−3940 ft*	*−4000 ft*	*−4050 ft*	*−4100 ft*
		Original Oil in Place (BSTB)		
Lower Clear Fork HFS* 2.2	0.2791	0.2791	0.2791	0.2791
Lower Clear Fork HFS* 2.1	0.5991	0.5995	0.5995	0.5995
Upper Wichita HFS* 2.0	0.2108	0.2170	0.2173	0.2173
Lower Wichita HFS* 2.0	0.2594	0.2746	0.2773	0.2779
Upper Wichita HFS* 1.2	0.1849	0.2063	0.2142	0.2164
Lower Wichita HFS* 1.2	0.0485	0.0737	0.0824	0.0907
Subtotal	1.5819	1.6501	1.6698	1.6810
Abo HFS* 1.1	0.0002	0.0281	0.1048	0.2357
Total	1.5821	1.6783	1.7746	1.9167

**HFS = high-frequency sequence.*

of water saturation is 90%. The OOIP value is pessimistically reduced when a 50% water saturation cutoff is used. Although intervals with less than 50% oil saturation can be considered nonproductive for waterflooding, the oil in these intervals is part of the OOIP. Nevertheless, a clear criteria for selecting a water saturation cutoff value do not exist. The tradeoff is that the recovery factor increases with decreasing OOIP.

Permeability Cutoffs

Three permeability cutoff values were considered: 0.1, 0.5, and 1 md. As shown in Figure 20, the equivalent porosity cutoff values for class I, II, and III rocks are 0.055, 0.07, and 0.09, respectively, when a 0.1-md permeability is used; 0.065, 0.09, and 0.12, respectively, when a 0.5-md permeability is used; 0.07, 0.10, and 0.14, respectively, when a 1.0-md permeability is used. Permeability cutoffs provide an easy way to determine rock-fabric-dependent porosity cutoff values.

The OOIP values for the entire FCU at an OWC of −3940 ft (−1201 m) are 1.537, 1.342, and 1.139 STB for 0.1-, 0.5-, and 1.0-md permeability cutoff values, respectively (Table 7). Figure 18D to F shows the change in pay zones with permeability cutoff, and Figure 19C shows that OOIP estimates decrease linearly with the permeability cutoff value. The selecting permeability cutoff value is field specific, depending on rock and fluid properties and operation considerations.

PRODUCTION-TREND ANALYSIS

Three-dimensional modeling is an iterative process involving the analysis and reanalysis of geology, geophysics, petrophysics, production data, and

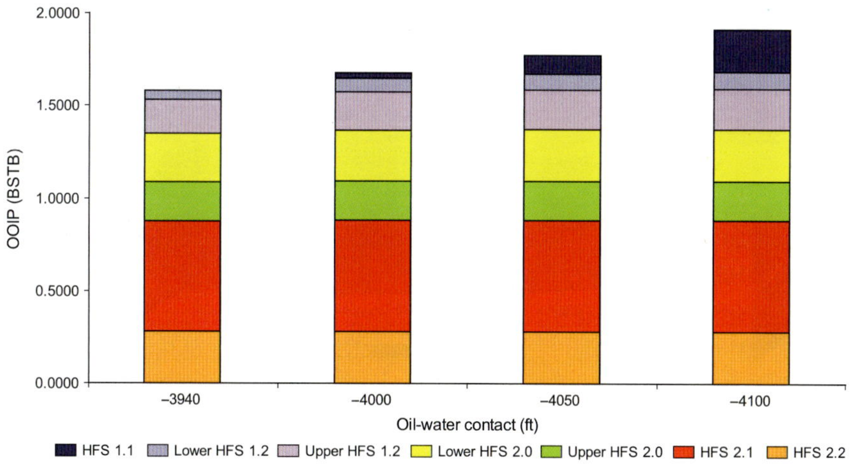

FIGURE 17. Effect of oil-water contact on original-oil-in-place (OOIP) estimate. HFS = high-frequency sequence.

Table 7. Effect of cutoff values on original oil in place.

Oil-Water Contact		*–3940 ft*		*–4100 ft Excluding Abo*	
Model Type		*Unit Model*	*Field Model*	*Unit Model*	*Field Model*
Cutoff Parameter	*Cutoff Value*	*Original Oil in Place (BSTB)*			
Porosity	0.03	1.577	1.688	1.676	1.789
	0.05	1.514	1.614	1.612	1.715
	0.07	1.344	1.423	1.440	1.521
	0.10	0.998	1.044	1.084	1.132
Water saturation	0.50	1.184	1.253	1.228	1.295
	0.60	1.323	1.406	1.378	1.460
	0.70	1.427	1.521	1.494	1.587
	0.80	1.505	1.608	1.584	1.686
	0.90	1.557	1.666	1.647	1.757
Permeability (md)	0.10	1.537	1.642	1.635	1.743
	0.50	1.342	1.434	1.433	1.528
	1.00	1.139	1.224	1.219	1.304

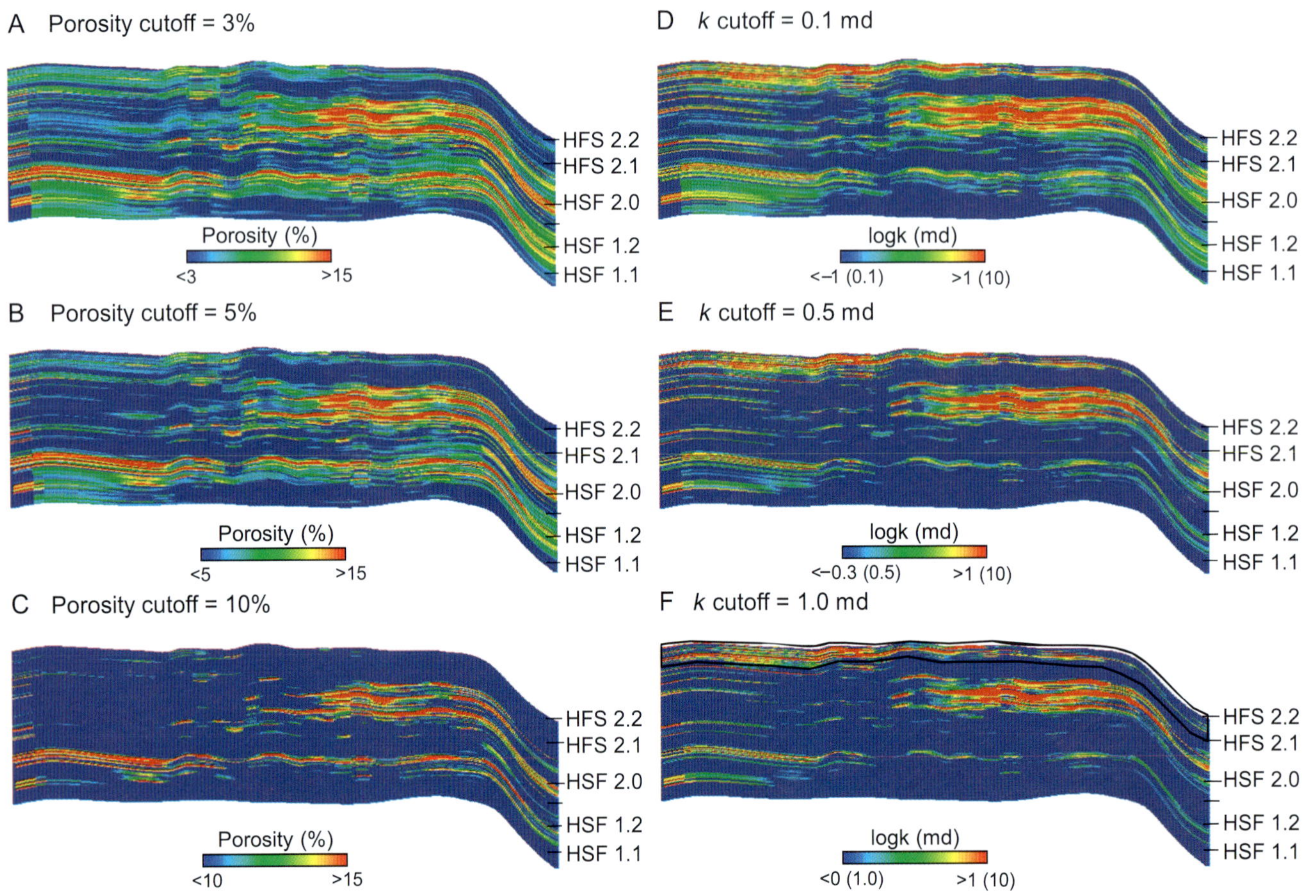

FIGURE 18. Effect of porosity and permeability (k) cutoff values on reservoir pay estimation. HFS = high-frequency sequence.

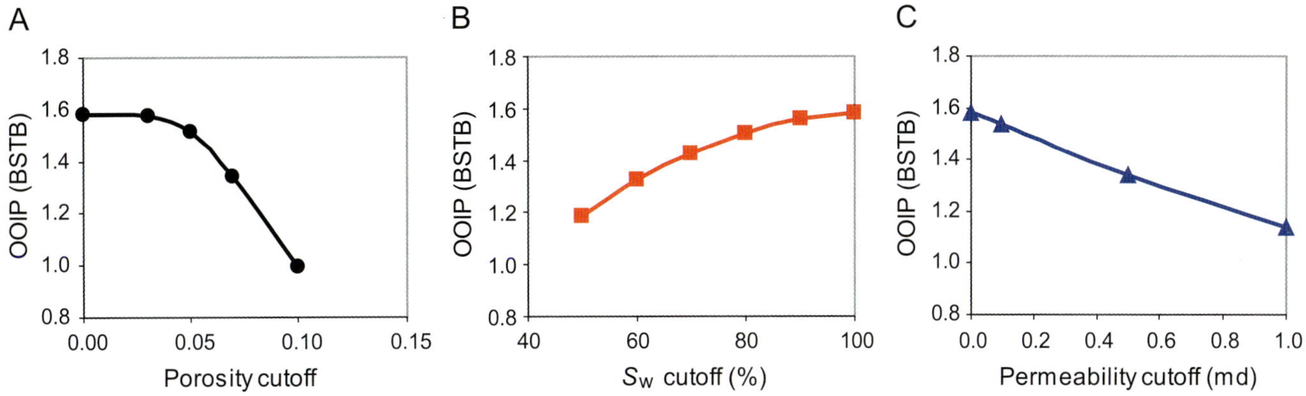

FIGURE 19. Effects of (A) porosity, (B) water saturation, and (C) permeability cutoff values on original-oil-in-place (OOIP) estimate. S_w = water saturation.

volumetric calculations. Problems with geologic or structural interpretation are normally detected during the construction of static models. Problems in petrophysical properties in well logs and static models can be identified using production data, production trends, and volumetrics.

Primary and waterflooding recovery data are shown as bubbles in Figure 21. During primary recovery, the production from the northern dome area (outlined by three ellipsoids) was significantly higher than in other parts of the field and was generally highest along a north-northwest–south-southeast trend (Figure 21A).

Oil recovery from waterflooding has been complicated by the use of widely varying injection patterns through time and episodes of infill drilling. Waterflooding operations have been dominated by 3:1 line-drive patterns (Figure 22) implemented in 1963. Oil recovery from such a pattern is expected to be higher at wells in the middle row (red circles) than at wells in the other two rows that are closer to the injectors (green circles). Four 3:1 line-drive patterns were created in the northern dome area during the initial phase of waterflooding (Figure 22). The expected higher recovery from the middle row of producers was observed in the three 3:1 line-drive patterns in the east, but not in the pattern in the west.

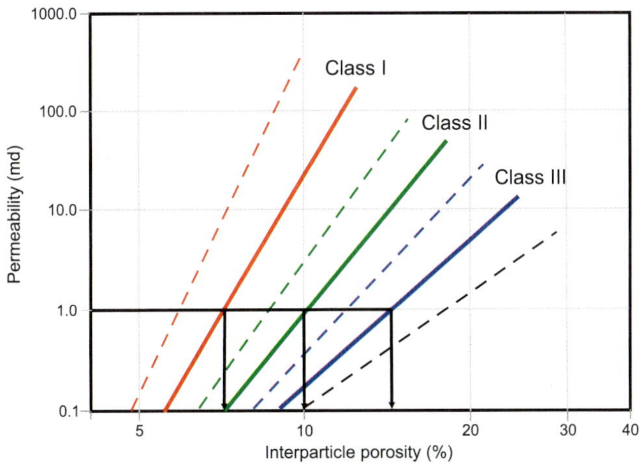

FIGURE 20. Relationship between permeability and porosity cutoff values. A specific permeability cutoff represents rock-fabric-dependent porosity cutoffs.

Figure 23 displays porosity, permeability, and $S_o\phi_h$ trends in the small-window study area in the northern dome part of the field. Figure 24 shows maps of the lower Lower Clear Fork and the upper Wichita in this same area. High production areas during primary recovery and waterflooding coincide well with high-porosity and $S_o\phi_h$ trends. Production from the central and eastern parts of this area has been higher because of high $S_o\phi_h$ values in both the Wichita and the Lower Clear Fork. Marked changes in petrophysical properties to the west have complicated recovery using patterned waterflooding in that area. Waterflooding recovery in this area decreases to the west (Figure 24).

REMAINING OIL AND OPPORTUNITIES FOR PRODUCTION ENHANCEMENT

The Fullerton Clear Fork field is a highly heterogeneous, thick reservoir with a low net-to-gross ratio and a low recovery efficiency, which translate as high remaining oil. With well spacing down to 10 to 20 ac, infill drilling cannot increase recovery substantially. Enhanced oil-recovery methods such as CO_2 flooding are needed to improve oil recovery by mobilizing residual oil.

In 1985, the operator (ExxonMobil) conducted a simulation study on an 800-ac CO_2 flooding test. Results suggest that miscible CO_2 flooding can triple oil

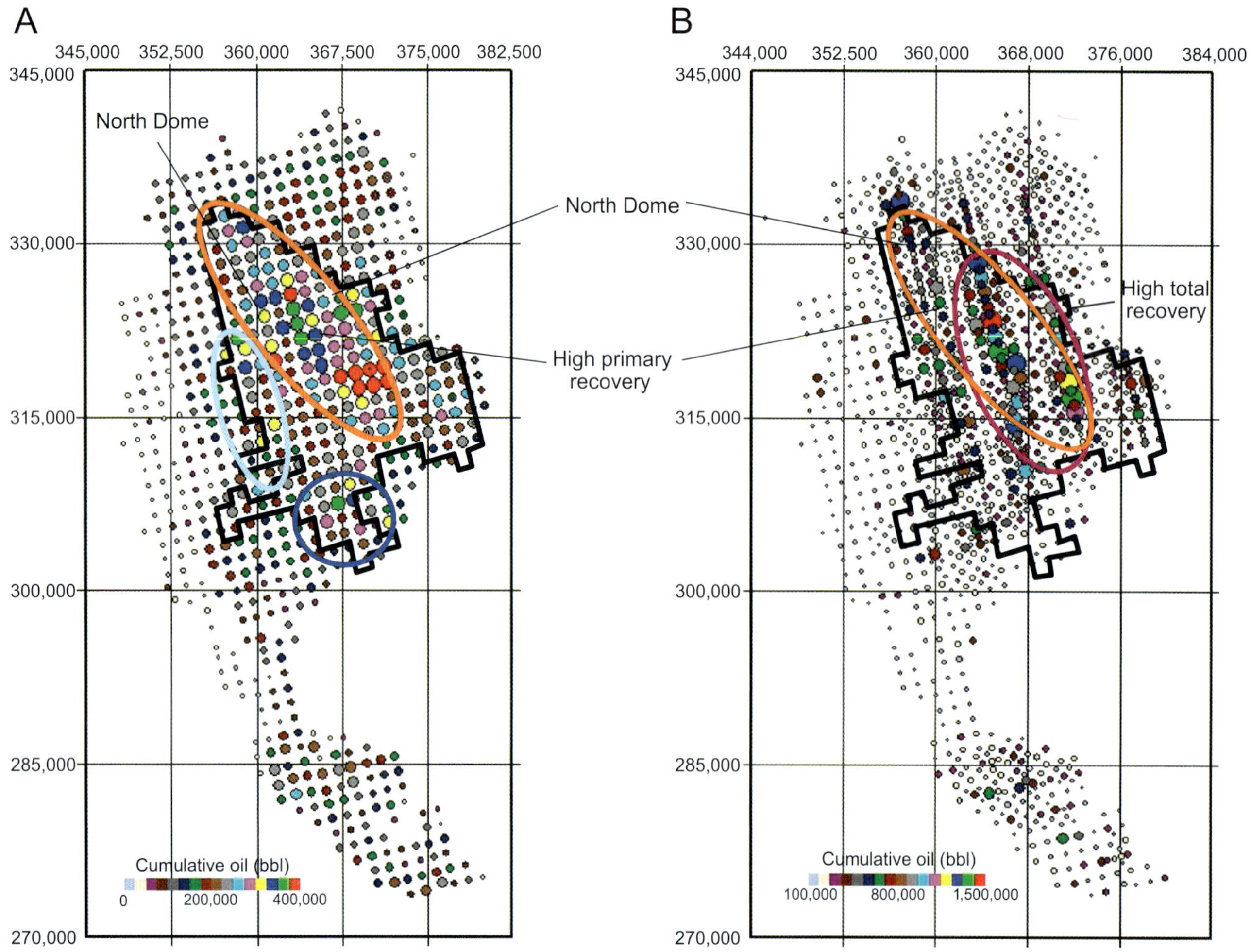

FIGURE 21. Production data and trends in (A) primary recovery and (B) total (primary and waterflooding) recoveries. Production is expected to be higher in the middle rows of producers (red) than the other two rows of producers (green) closer to the injectors (green).

production and improve from 30 (from waterflooding) to 50%. Despite this optimistic forecast, note that the 800 ft (244 m) reservoir column in the FCU is very thick and very complex to be successfully flooded with CO_2. A more viable alternative is to perform area- and interval-selective CO_2 flooding (Wang et al., 2007). The middle and lower Wichita reservoirs with good porosity but low matrix permeability and unpredictable karst-related breccias are very risky for CO_2 flooding. Recovery from CO_2 flooding can be correlated to waterflooding efficiency. As a rule, the better the waterflooding response, the better the CO_2 flood recovery will be. More than half of the OOIP was located in the lower Lower Clear Fork, and the upper Wichita (Figure 15) contains most of this in the eastern part of the north dome, which has the best reservoir quality (Figure 24). The waterflooding recovery efficiency has been good in this area, and the potential for CO_2 flooding is high.

RESERVOIR SIMULATION AND SENSITIVITY ANALYSIS

Because reservoir simulation was performed only on the small-window model area, the objectives were to investigate critical geologic and engineering factors affecting production performance and to locate remaining oil.

Reservoir Model Construction

To develop an early understanding of the geology and petrophysics in the 2000-ac study area, a 2000-ac area (Figure 22) containing the highest density of cores and the best suites of wireline logs was selected for detailed, integrated, geologic, petrophysical, and engineering characterization. In the fieldwide model, this area extends from the highly productive region in the east to the poor-quality reservoir in the west. As shown in Table 2, only 20 HFCs were identified

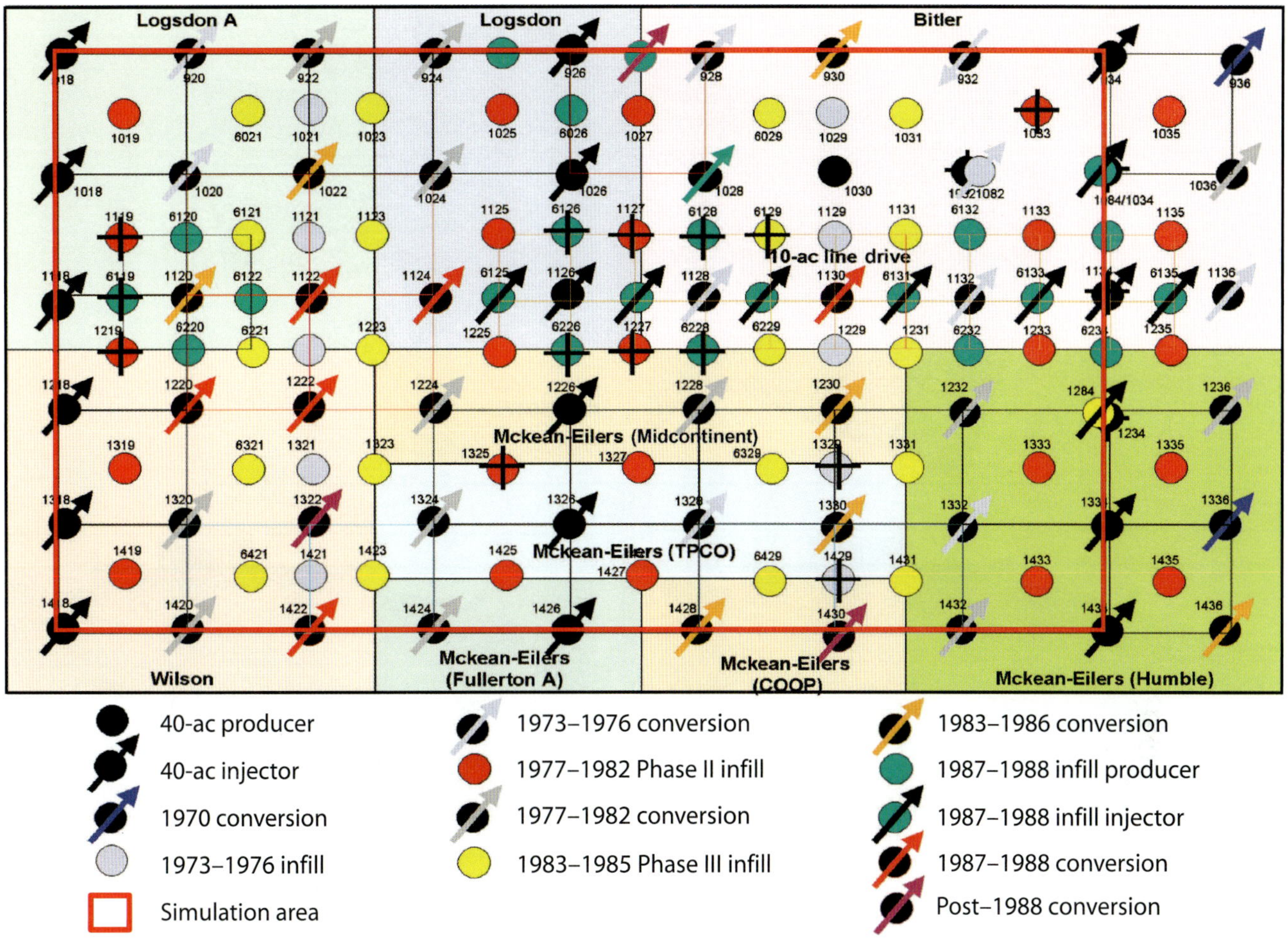

FIGURE 22. **Location and water injection patterns in the 2000-ac pilot area. The area of reservoir simulation is indicated by the red box.**

from core and wireline-log studies in the Lower Clear Fork and Wichita formations during the study period. Each cycle was further divided into upper and lower flow units, which were used to provide a geologically constrained framework for 3-D modeling.

Flow-Unit Modeling

Distributing petrophysical properties within a geologic model is one of the most difficult tasks in building a simulation model. In this study, as well as in other reservoir studies done at the Bureau of Economic Geology, the basic geologic framework within which the properties are distributed is the HFC. Outcrop studies, however, have shown a systematic layering of petrophysical properties within the HFC related to the vertical succession of rock fabrics (Kerans et al., 1994; Lucia et al., 1995). High-frequency cycles tend to show an upward increase in carbonate grains, resulting in an ideal vertical succession from mud-dominated fabrics (mudstone, wackestone, and mud-dominated packstone) to grain-dominated packstone and grainstone. Each fabric tends to form a geologic bed, and each fabric has a specific porosity-permeability transform. Therefore, each rock-fabric bed can be defined as a layer within the HFC.

Values for porosity, permeability, and initial water saturation are calculated from wireline logs and provide the basic data used to popularize the geologic model. Numerous statistical methods for filling the interwell space with the calculated properties exist. Whatever method is used, it must be constrained by the rock-fabric flow layers to preserve the petrophysical heterogeneity. The rock-fabric approach for distributing petrophysical properties is illustrated by outcrop results (Figure 25) (Lucia et al., 1992; Lucia, 1999). At Lawyer Canyon, Guadalupe Mountains, New Mexico, a single HFC has been mapped for 2700 ft (823 m) along an outcrop, and about 2000 permeability measurements were made. Statistical analysis of the permeability data show 1) a significant difference

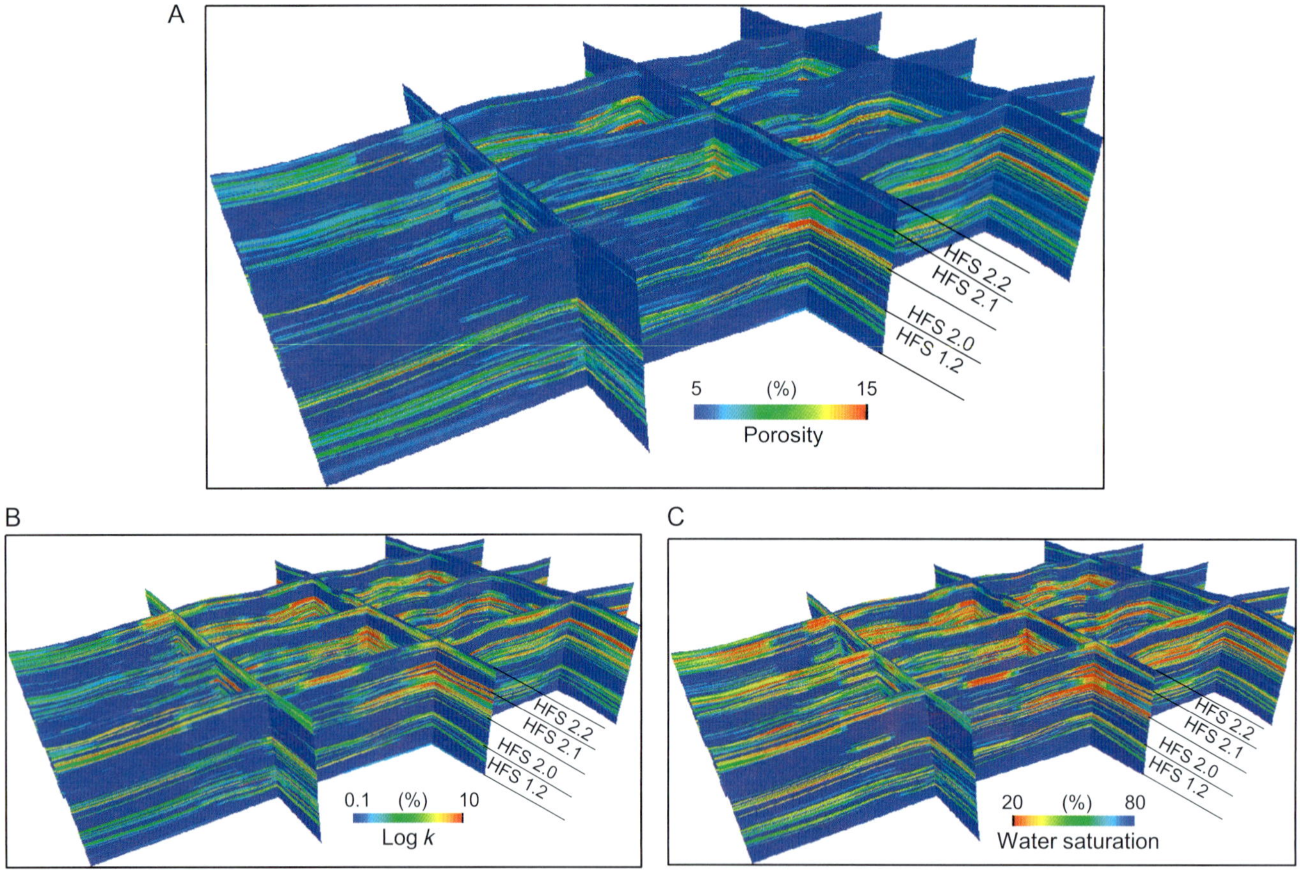

FIGURE 23. Petrophysical trends in (A) porosity, (B) permeability, and (C) water saturation distributions in the 2000-ac small window area. HFS = high-frequency sequence.

between the average permeability in the upper grainstone and the lower mud-dominated fabrics and 2) little or no spatial correlation of the permeability values within a rock-fabric layer, in this case, the grainstone layer (Lucia et al., 1992). The implications are that only rock-fabric layers can be correlated between wells, not petrophysical properties, and that petrophysical properties should be statistically distributed within rock-fabric layers, not within HFCs.

In this study, the HFCs are correlated using porosity logs calibrated to core descriptions (Ruppel and Harrington, 2012). Cycles within the Lower

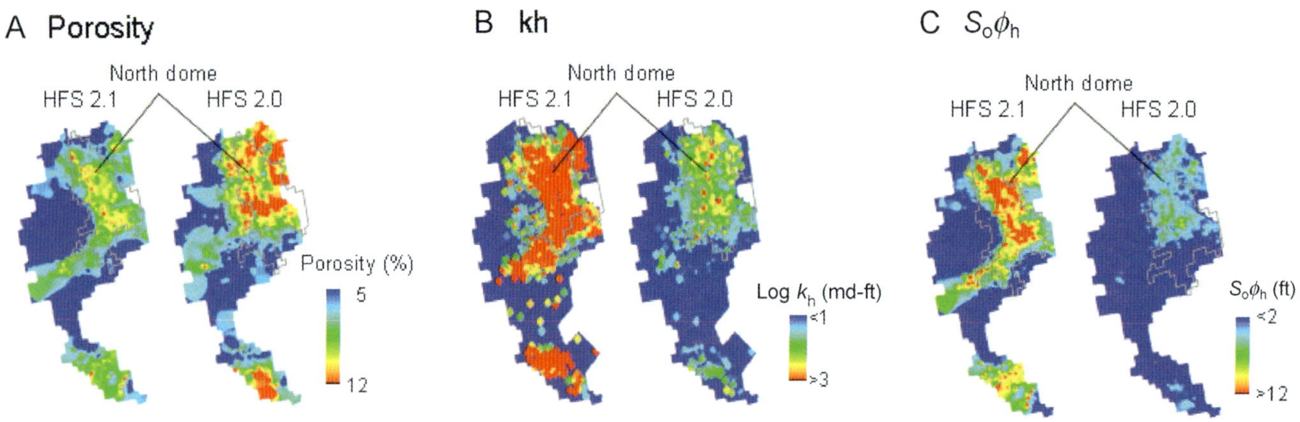

FIGURE 24. Maps of potential CO_2 flooding targets in the Lower Clear Fork (HFS 2.1) and upper Wichita (HFS 2.0). The areas with good reservoir quality, hydrocarbon volume, and production are excellent targets. HFS = high-frequency sequence.

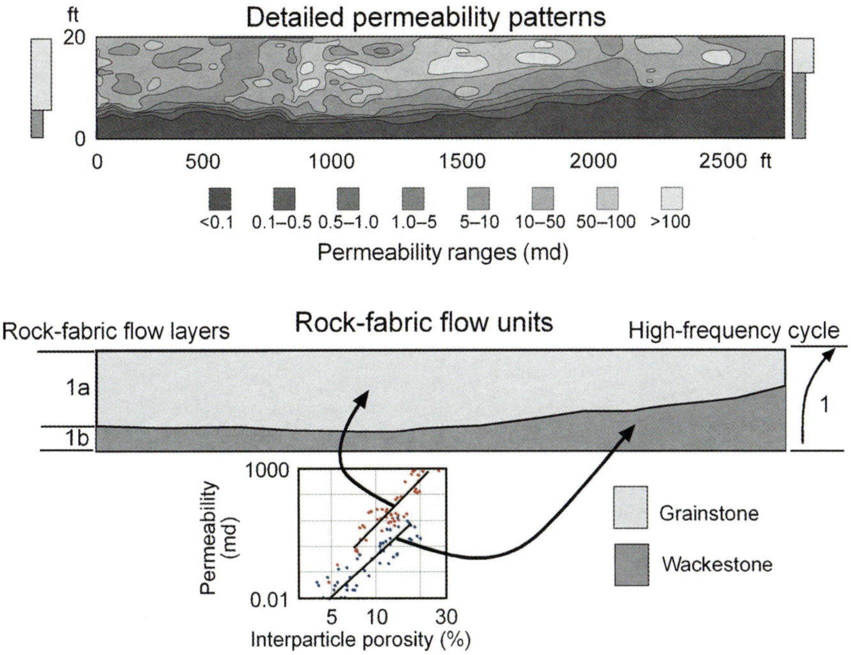

FIGURE 25. Lawyer Canyon cycle 1 flow-layer model.

Clear Fork sequences L2.2 and L2.1 are defined by an upward increase in graininess. Wichita cycles are composed of mud-dominated fabrics and are typically capped by tidal-flat fabrics. The cycles are typically dolostone, but limestone may be present at the cycle base (Ruppel and Harrington, 2012).

The grainy top of the Lower Clear Fork HFCs tends to be more porous than the muddy base, which is true for both limestones and dolostones. As pointed out by Ruppel and Harrington (2012), much of the core-log calibration was done by comparing core descriptions with porosity logs. The calibration can also be done by comparing thin-section descriptions with porosity measurements when the thin sections are taken directly from the core plugs. An analysis of rock fabrics and porosity using data from sequences L2.2 and L2.1 (Figure 26) shows that, for dolostones, grain-dominated packstones average 11.5% porosity; mud-dominated packstones, 9.4%; and wackestones and mudstones, 4.9%. For limestones, grainstones average 14.9% porosity; grain-dominated packstones, 8.2%; mud-dominated packstones, 7.6%; and wackestone and mudstones, 7.0%. Although a significant overlap in the porosity ranges exists, a clear increase in porosity with increasing graininess in the dolostones is observed. In the limestones, the grainstones (moldic grainstones) are clearly more porous than the other fabrics. The lime grain-dominated packstones have an average porosity similar to the mud-dominated fabrics. This result may be related to the difficulties in distinguishing between grain- and mud-dominated fabrics. The distinction commonly becomes a matter of judgment by the observer as to whether the intergrain volume is completely or partially filled with mud.

For simplicity, the HFCs are divided into two rock-fabric flow layers, a lower low-porosity layer and an upper high-porosity layer. In Lower Clear Fork sequences L2.1 and L2.2, the porosity changes are used as a surrogate for a lower mud-dominated fabric and an upper grain-dominated fabric. Because a considerable overlap of porosity values for each fabric exists, a single porosity value cannot be used to separate rock fabrics. Instead, the vertical porosity profile must be used and a judgment must be made as to where a significant change in rock fabric is likely to occur. Seven HFCs are defined in sequence L2.1 with two flow layers each except for cycle 2105, which did not appear to have a low to high porosity profile in the simulation area (Figure 27). The cycle types correlate quite well with porosity, grain-dominated fabrics, and tidal-flat facies, but not with lithology. In most cycles, the distinction between low porosity and high porosity is quite clear. The upper two cycles are class 3 fine-crystalline dolostones and the porosity is very low to be productive.

Three HFCs were originally defined in sequence L2.2 and are illustrated in Figure 28. The cycles were defined based on limestone upward-shoaling successions found in the core from well 6122 and were difficult to pick in the dolostone wells because of low porosity values. As such, the flow layers in this

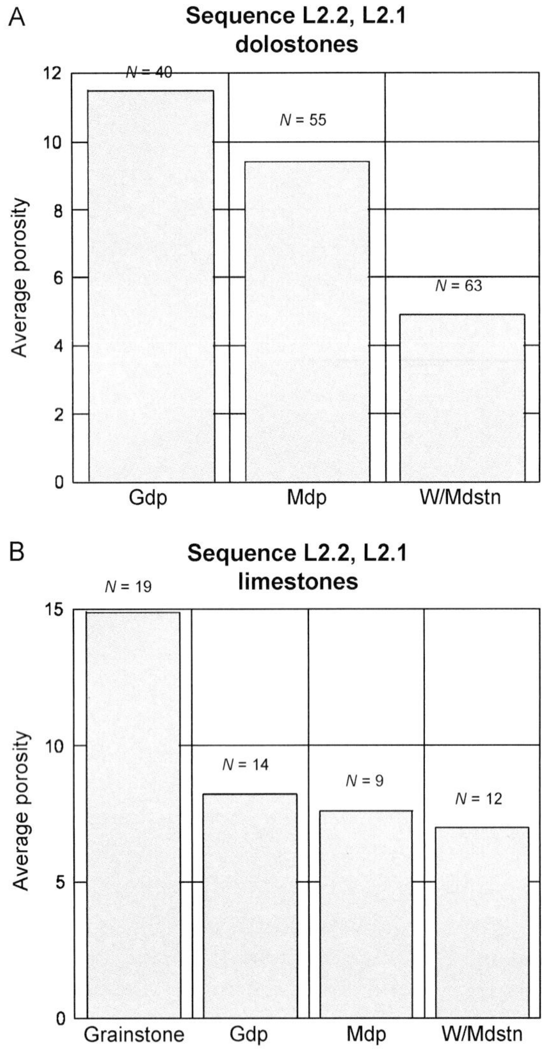

FIGURE 26. Average porosity for various rock fabrics in sequences L2.1 and L2.2. (A) Dolostone rock fabrics showing a decrease in porosity with increasing lime mud. (B) Limestone rock fabrics showing high porosity correlated with grainstones. Grain-dominated packstones may have sufficient lime mud to be described as mud-dominated packstones. Gdp = grain-dominated packstones; n = number of samples; Mdp = mud-dominated packstones; W/Mdstn = wackestones and mudstones.

sequence are somewhat arbitrary. Sequence L2.2 was later divided into eight HFCs based on core descriptions outside of the simulation area (Ruppel and Harrington, 2012). The simulation model has six flow layers, whereas the full-field model has eight HFCs. An imperfect match between the flow layers in the simulation model and the new HFCs (Figure 28) exists. An attempt to integrate the eight HFCs into the simulation model has not been made.

In the Wichita, a clear relationship between fabric and porosity does not exist because the cycles are typically composed of mud-dominated fabrics, either limestone or dolostone. The limestones, however, have less than 3% porosity and are typically found at the base of the cycles that have porous dolostone tops. However, many cycles have no limestone, and the upper intervals are more porous than the lower intervals. Ten HFCs were picked in the Wichita with two flow layers each (Figure 29). The upper cycles from W1 to W5 typically have thin limestones at the base and are easily correlated throughout the simulation area. No cycles were picked between cycles W5 and W8 because the interval is composed of limestone and dolostone, typically has less than 10% porosity, and therefore is nonproductive. Cycles W8 to W12 are based solely on the porosity profiles and have little relationship to tidal-flat-capped cycles or lithology. Therefore, these cycles and flow layers are the most arbitrary of all the layers in the simulation model.

The importance of the flow layers is to preserve layers of high and low permeability (Figure 30). The individual cycles, however, are characterized by a single transform. The Wichita is characterized by a class 3 transform because it is typically a lime wackestone or fine-crystalline dolowackestone. The Lower Clear Fork L2.1 sequence is characterized by a class 2 transform in the lower interval and a class 3 transform in the upper interval. The Lower Clear Fork L2.2 sequence is characterized by a class 1 transform. Therefore, the permeability contrast between the upper and lower flow layers is related to the difference in porosity, not to the difference in rock fabric. A more accurate permeability contrast could be obtained if the different limestone fabrics could be distinguished.

Flow Modeling

Of the total 140 wells in the small-window model area (Figure 22), 85 were initially selected for detailed geologic and petrophysical characterization. However, early 3-D models based on these 85 wells revealed that the horizons mapped using only these wells did not match well tops at unselected wells and that boundary-effect problems at the edges of the model were significant. To correct these problems, all wells in the study area and a line of wells immediately outside the study area were subsequently correlated.

A fine-scale geologic model containing 3.2 million cells (Table 3) was then constructed for the area, and the porosity was mapped through the 3-D space (Figure 23). Permeability and water saturation (Figure 23B, C) were calculated from porosity (Figure 23A) on the basis of correlations provided by Lucia (2007). Rock-fabric-dependent permeability-porosity relationships are shown in Figure 20. The

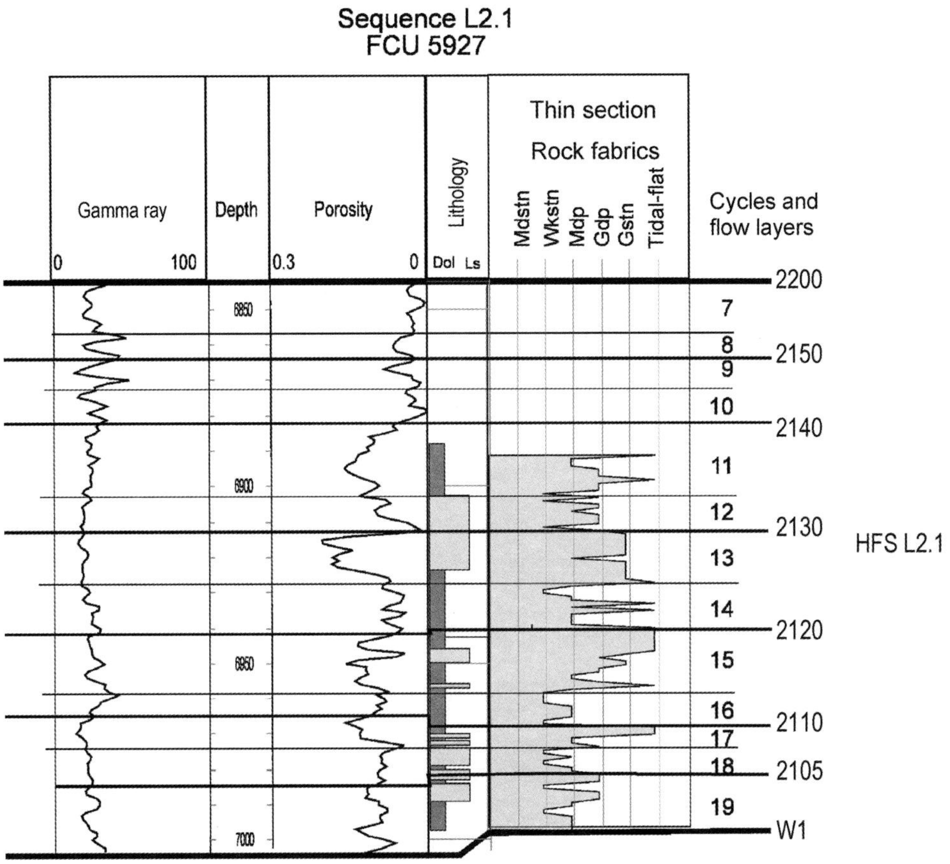

FIGURE 27. Flow layers and cycles in sequence L2.1. FCU = Fullerton Clear Fork Unit; Dol = dolostone; Ls = limestone; Mdstn = mudstone; Wkstn = wackestone; Mdp = mud-dominated packestone; Gdp = grain-dominated packestone; Gstn = grainstone.

rock-fabric-based water saturation-porosity relationships used in the initial 2000-ac model differ from those used in the full-field model.

A careful examination of Figure 24A and B shows that, although the ϕ_h in the model area is higher in the Wichita, k_h is greater in the Lower Clear Fork because of the different rock fabrics in each. Note that oil saturation (Figure 24C) is highest in the Lower Clear Fork, again because of the presence of better quality rock fabrics in this part of the reservoir.

Once the fine-scale reservoir model was constructed, it was scaled up to a coarse model containing 130,000 cells (73 × 48 × 39) for reservoir simulation. Figure 31 shows porosity and initial oil saturation in the upscaled flow model. In Figure 31A, the reservoir appears compartmentalized both laterally and vertically because only porosity values of greater than 6% are displayed.

Reservoir Simulation and Sensitivity Analysis

The simulation study was divided into two phases: sensitivity analysis and history matching. From the sensitivity study, it is possible to rank the importance of reservoir parameters affecting production performance.

Figure 32 displays simulated oil distributions in 1942, 1970, 1980, and 2002. Water injection was initiated in 1960. The blue areas in the 1960 image represent high gas saturation. As water injection progressed, water displaced oil from the injectors into these high gas-saturation areas, as seen in the 1970 and 1980 images.

Figure 33 shows simulated oil-saturation distributions in 2002 in the two model layers of the Lower Clear Fork HFS L2.2. Because this interval is highly heterogeneous with poor reservoir continuities between injectors and producers, the sweep efficiency during waterflooding has been low.

Overall, our modeling shows that the Lower Clear Fork HFS L2.1 contains the highest ϕ_h, k_h, and oil in place, especially in the highest porosity areas in the eastern part of field (Figure 24). However, the reservoir continuity in this area is somewhat limited, as revealed by the presence of many small areas of low oil saturation. Simulated oil-saturation distributions in 2002 at layers 11, 13, 15, and 17 illustrate this

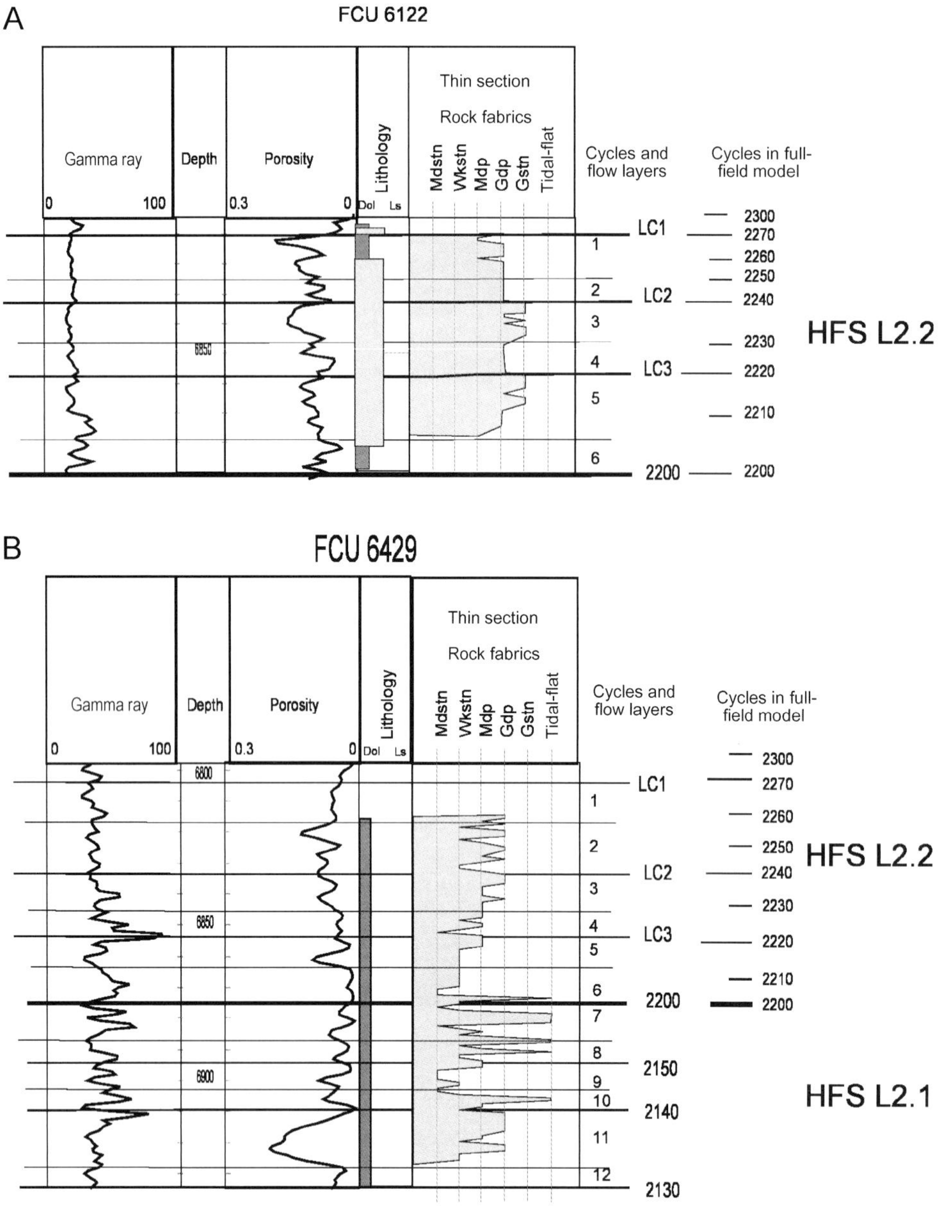

FIGURE 28. Flow layers in sequence L2.2. The cycles and flow layers in the simulation model are compared with the cycles in the full-field model. (A) In the limestone area, a reasonable correlation between cycles, flow layers, and porosity is observed. (B) In the dolostone area, the correlation between cycles, flow layers, and porosity is less apparent. FCU = Fullerton Clear Fork Unit; HFS = high-frequency sequence; Dol = dolostone; Ls = limestone; Mdstn = mudstone; Wkstn = wackestone; Mdp = mud-dominated packestone; Gdp = grain-dominated packestone; Gstn = grainstone.

heterogeneity and poor waterflood sweep efficiency (Figure 28).

In the model area, porosity and reservoir continuity are best developed in Wichita cycles (L2.0; W1–W3). Simulated oil-saturation distributions (Figure 35) in 2002 of two layers, 20 and 22 (cycle top flow units in L2.0 cycles W1 and W2), indicate that these intervals have been fully swept except for the southeasternmost corner of the model area, where the upper Wichita interval is not perforated in several wells. The lower upper Wichita in this area exhibits low permeability. This low-permeability interval separates porosity and oil saturation in the uppermost Wichita from relatively high porosity, permeability, and oil saturation in the lower Wichita. Simulated oil-saturation distributions in 2002 in model layers 30 and 32 (Figure 36) indicate that Wichita layers W8 and W9 have been well swept in areas where wells have been perforated. The southeasternmost corner of the model area, however, has not been flooded because many injectors in this area have not been perforated.

Sensitivity Study

The reservoir flow model was used for the history matching of production data of the simulation window (Figure 22). History matching was complicated by lack of reliable primary production data, gas injection, upscaling, relative permeability, and karst breccias and fractures.

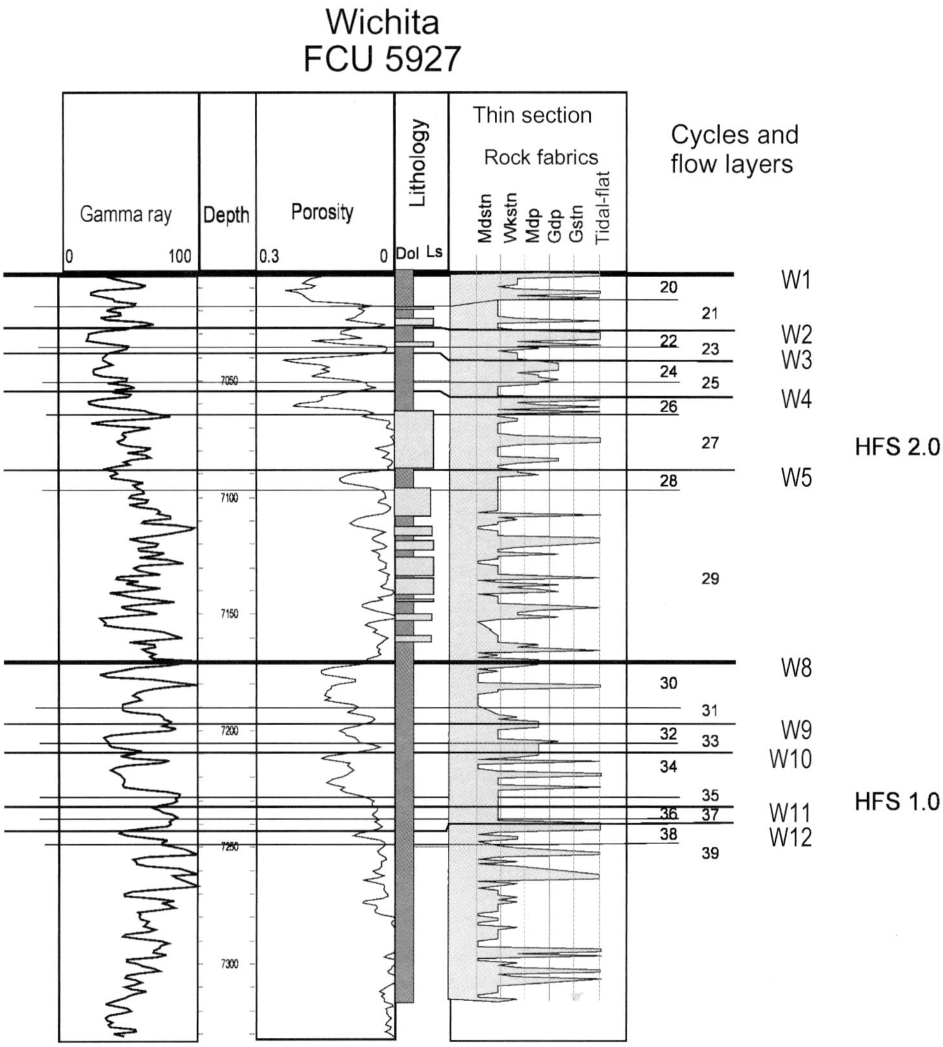

FIGURE 29. Flow layers in the Wichita. FCU = Fullerton Clear Fork Unit; Dol = dolostone; Ls = limestone; Mdstn = mudstone; Wkstn = wackestone; Mdp = mud-dominated packestone; Gdp = grain-dominated packestone; Gstn = grainstone.

The issues of upscaling from a fine-grid to a coarse-grid model tend to average out the permeability streaks, which in turn underestimate fluid channeling. Upscaling affects absolute permeability, relative permeability, and the ratio of horizontal and vertical permeabilities (k_{vh}). Relative permeability and the ratio of k_{vh}, two of the most dominating factors affecting oil recovery from reservoir simulation, need to be optimized through history matching.

Six relative permeability curves for gas, oil, and water phases in drainage and inhibition processes are observed. They raise the question of what the most critical relative permeability is. A good match with primary recovery is essential for a good match with secondary and tertiary recoveries. A sensitivity study of oil relative permeability for primary recovery was performed by changing Corey's exponent of oil curve from 2 to 5 because oil permeability and oil recovery decrease with an increase in Corey's exponent (Figure 37). Note that oil permeability had no effect on primary recovery from 1942 to 1960 but did have a strong effect on waterflooding. Oil permeability in primary recovery affects pressure instead of volumetrics. The waterflooding history cannot be matched if the pressure history during primary recovery has not been matched, and it is crucial to use proper oil relative permeability curves for primary recovery.

One of the difficulties in this simulation was to match water production and injection rates. Both water production and injection from the area are significantly higher than those simulated on the basis of matrix permeability in the static model. Nonmatrix flow has been indicated by negative skin factors from −3 to −5 in well-testing data, anomalous water production, and breccias reported from the Wichita in several cores in the field (Ruppel and Harrington, 2012). In addition, near-wellbore fractures might have been induced by high-pressure water injection. Hence, a negative skin factor of −3

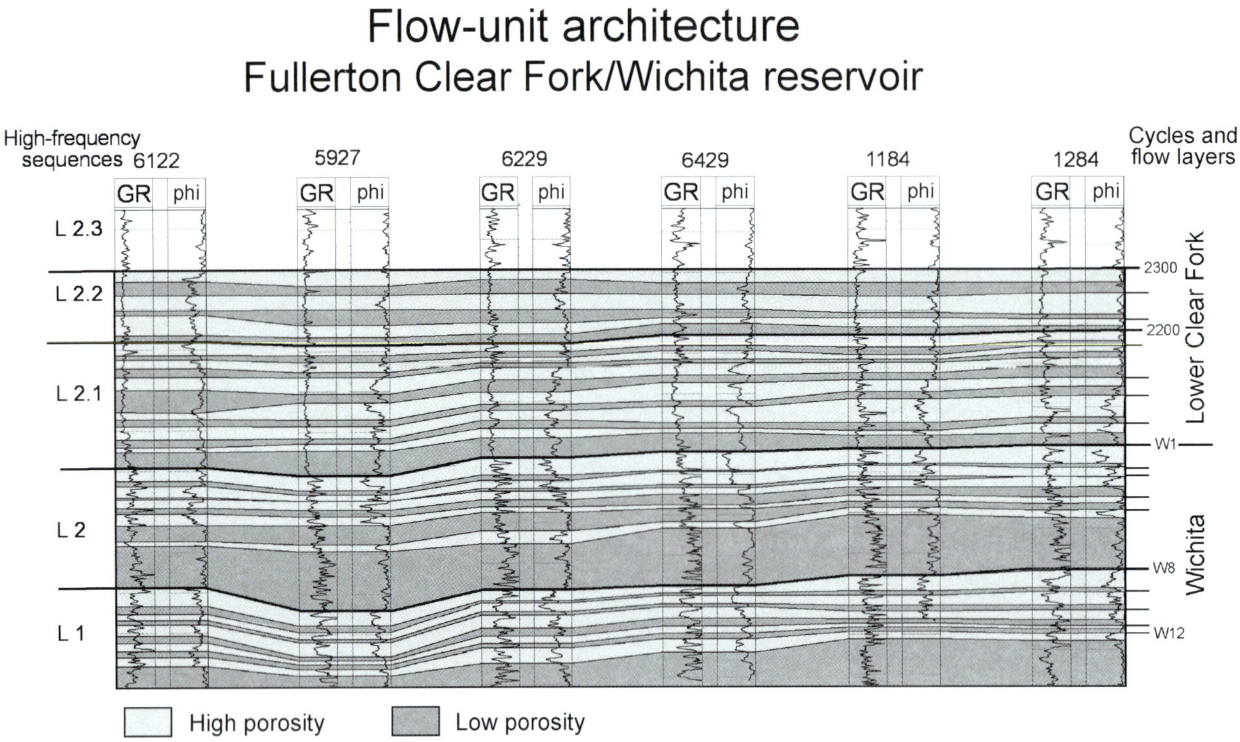

FIGURE 30. Lower Clear Fork and Wichita flow layers in cored wells of the simulation area illustrating the use of porosity to build a flow model. GR = gamma ray; phi = porosity.

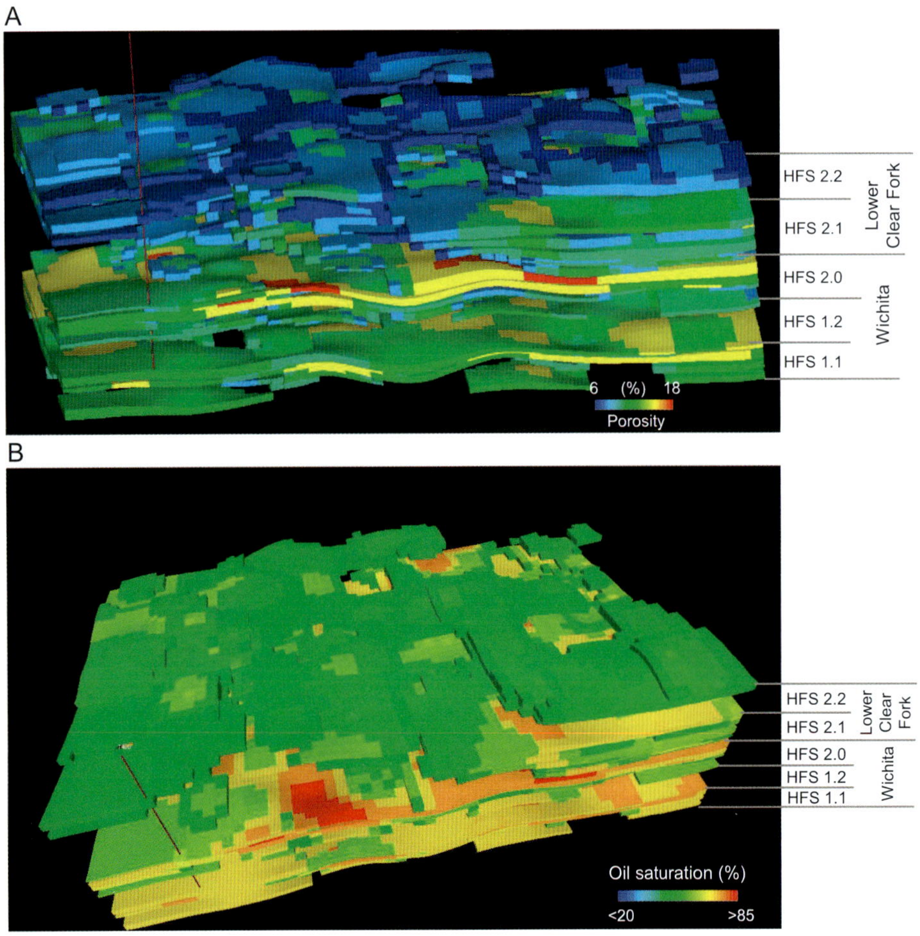

FIGURE 31. (A) Porosity and (B) initial oil saturation in the upscaled simulation model, small window area. HFS = high-frequency sequence.

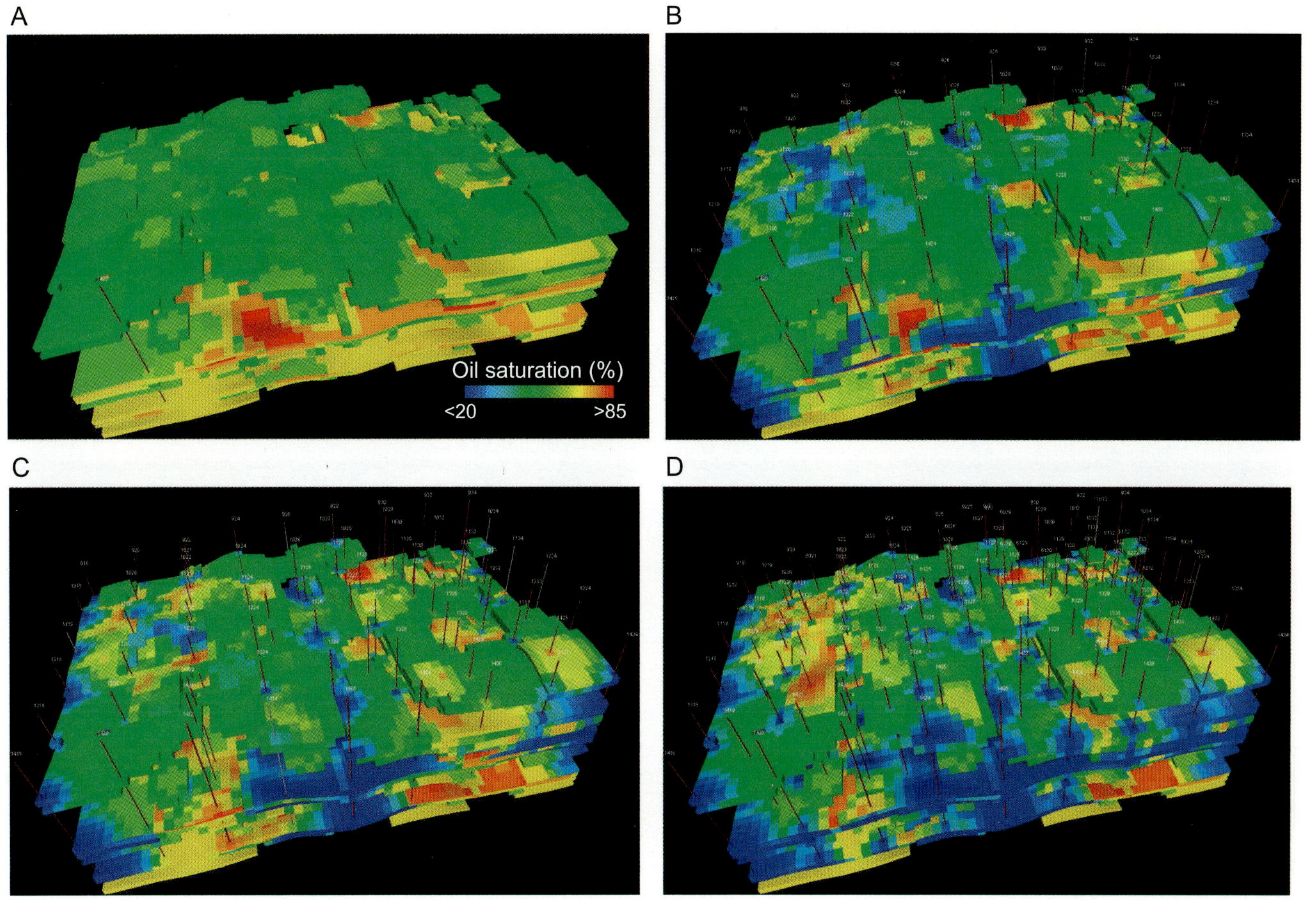

FIGURE 32. Simulated oil saturations in (A) 1960, (B) 1970, (C)1980, and (D) 2002 in small-window model area.

(or effective wellbore radii) was used to simulate near-wellbore fractures, and permeability values in the lower Wichita were modified to simulate karst-related breccias. Two simulation runs were performed by increasing the permeability values in the lower Wichita by factors of 25 and 30. Figure 38 compares the results of these simulations with cumulative water production (black dashes). Without permeability modification, simulated water production (blue curve in Figure 38) is significantly lower than field data (crosses). Increasing the permeability values in the lower Wichita by a factor of 25 (green curve) and 30 (red curve) significantly improves the matches between simulated water production

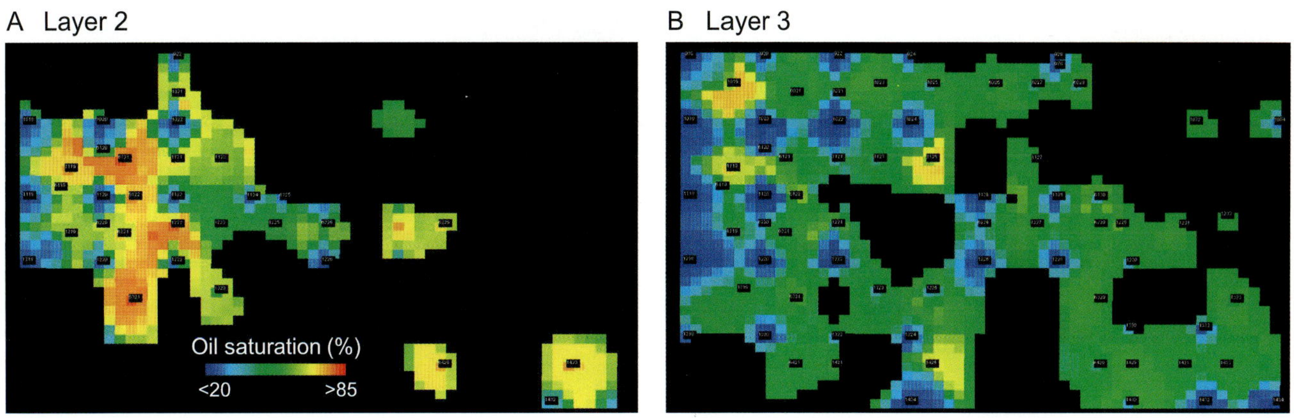

FIGURE 33. Simulated oil saturations in 2002, in upper cycles of the Lower Clear Fork high-frequency sequence (HFS) 2.2 in the small-window model area. See Table 2 for the relationship between model layers and sequences and cycles.

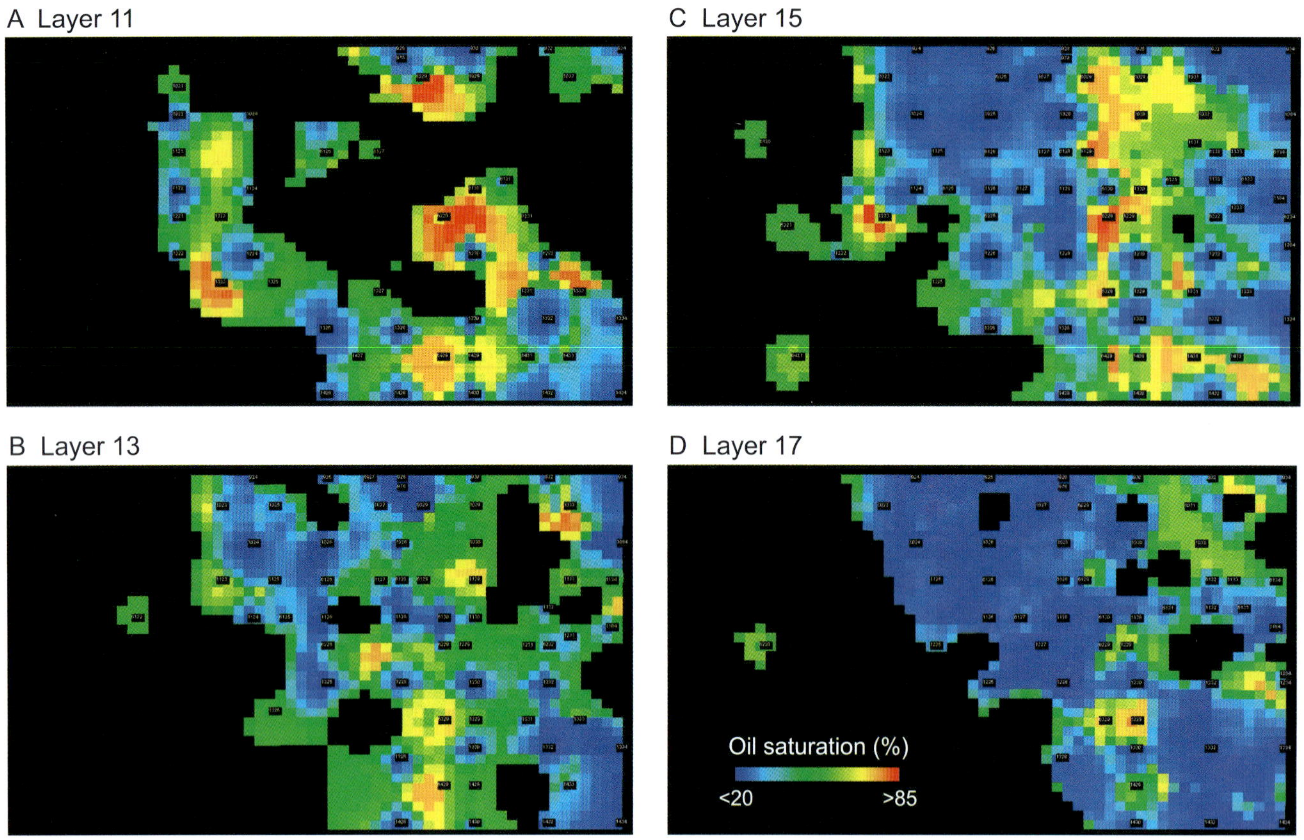

FIGURE 34. Simulated oil saturation in 2002, in cycle-top flow unit layers in Lower Clear Fork high-frequency sequence (HFS) 2.1 in the small-window model area. See Table 2 for the relationship between model layers and sequences 2 and cycles.

and field data. However, in both cases, the higher permeability values tend to overestimate water production during the early stages of waterflooding (1960–1985).

Simulation results provide an important glimpse into reservoir response in west Texas Clear Fork carbonate reservoirs. Both techniques were used to construct the model and the results and can serve as a basis for

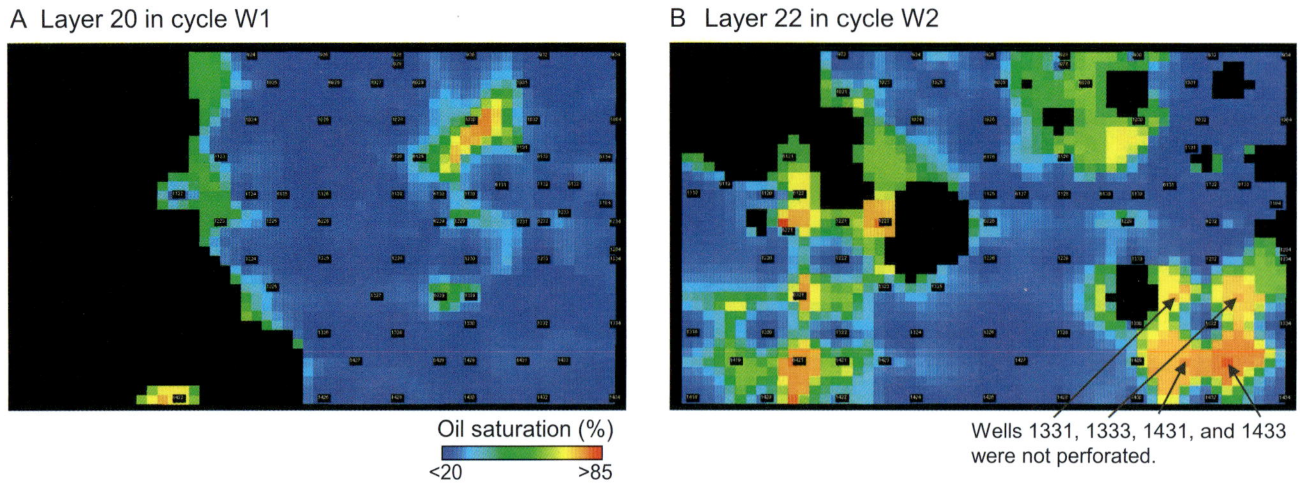

FIGURE 35. Simulated oil saturation in 2002, in the upper part of the Wichita (high-frequency sequence [HFS] 2.0) in the small-window model area. See Table 2 for the relationship between model layers and sequences and cycles.

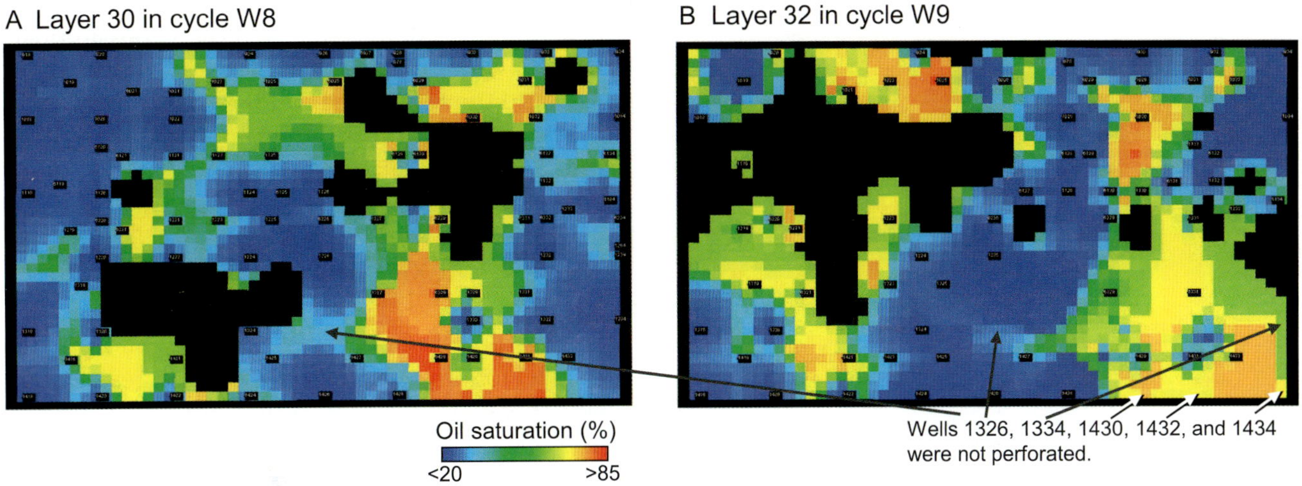

FIGURE 36. Simulated oil saturation in 2002 in model layers of the middle Wichita (high-frequency sequence [HFS] 1.2 and HFS 2.0) in the small-window model area. Note that the oil has not been produced or swept from the southeasternmost part of the model area.

improved modeling and simulation of other shallow-water carbonate-platform reservoirs.

SUMMARY

The Fullerton Clear Fork field is a highly heterogeneous reservoir with complex lithology and porosity trends controlled by original depositional facies and diagenetic overprint. Porosity is better developed in the mud-dominated inner-ramp Wichita Formation than in the overlying subtidal Lower Clear Fork Formation. Although ϕ_h is higher in the Wichita, hydrocarbon volume and k_h are greater in the Lower Clear Fork because of differing rock fabrics.

In HFS 2.2, reservoir continuity and quality are best in the southern dome area and generally poor in the north dome, with scattered local porosity developments. With the highest hydrocarbon pore volume and good ϕ_h and k_h, the high-porosity trends in the Lower Clear Fork and the upper Wichita are the high reservoir-quality units in the Fullerton Clear Fork field. Reservoir quality decreases sharply away from these porosity trends. The middle W4 and W5 display low permeability in the northern half of the north dome and become porous in the southern half of the north dome and the south dome. The underlying middle Wichita reservoir is a more homogeneous unit with high porosity and continuity. However, karst-related breccias and fractures have complicated production and injection in this interval.

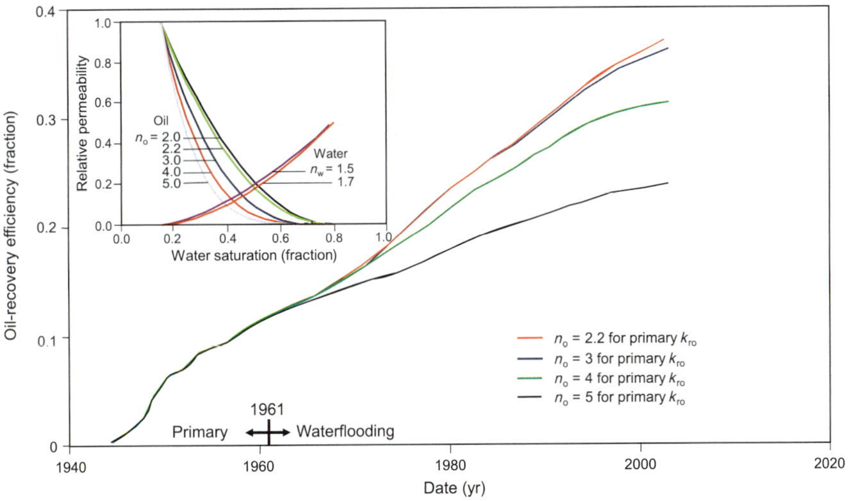

FIGURE 37. Effect of oil-water relative permeability on predictions of waterflooding recovery in the Fullerton Clear Fork reservoir. Oil recovery increases with the increase in oil relative permeability, which decreases with the Corey exponent (n_o).

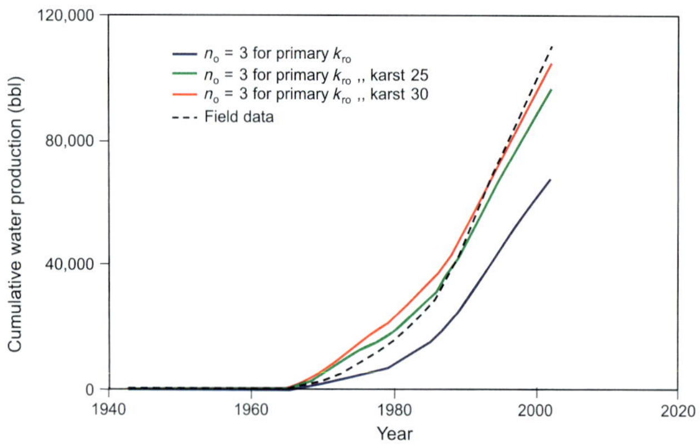

FIGURE 38. Effect of permeability multiplication on water production. Because karst intervals and water-injection–induced fractures are common in Fullerton, water production (dashed curve) has been much faster than predicted (blue curve), and post-1980 water production can be better matched with those simulated with an increase of permeability by a factor of 25 (green curve) to 30 (red curve).

Original oil in place is a complex function of log-data quality, mapping parameters, vertical resolution of 3-D grid, OWC contact, and cutoff values in porosity, permeability, and water saturation. Permeability and water saturation, equivalent to rock-fabric–dependent porosity cutoffs, can simplify the criteria of cutoff values. High vertical-resolution 3-D models should be used in volumetric calculation. The OOIP estimate depends strongly on the selection of cutoff values. In the case of OWC at −3940 ft (−1201 m), the estimated OOIP is 1.58 BSTB without using any cutoff, 1.32 BSTB with a 5% porosity cutoff, 1.23 BSTB with a 0.5-md permeability cutoff, and 1.32 BSTB with a 60% water saturation cutoff.

In simulation, oil relative permeability for primary recovery has a strong effect on recovery from waterflooding. To match the waterflooding history, both production and pressure histories in primary recovery must be matched. Because most high-permeability streaks were averaged out by upscaling, both oil and water relative permeabilities used in simulations were higher than those measured in the laboratory.

Because fractures and breccias are common in testing and core data, negative skin factors (or effective wellbore radii) were used to simulate near-wellbore fractures, and permeability values in the lower Wichita were modified to simulate karst-related breccias. Optimal fluid and rock properties were determined through history matching, and simulation results provide an important glimpse into reservoir response in west Texas Clear Fork carbonate reservoirs. Both the techniques used to construct the model and the results and can serve as a basis for improved modeling and simulation of other shallow-water carbonate-platform reservoirs.

ACKNOWLEDGMENTS

We thank Julia Gale, James Jennings, and Stephen C. Ruppel for providing extensive critical review comments that greatly improved the content and quality of the manuscript. Terry Anthony and David Smith, from ExxonMobil Corporation, openly shared their knowledge of field operations and data and offered valuable feedback on modeling results. Lana Dieterich edited the manuscript. The illustrations were prepared with the assistance of the graphics group at the Bureau of Economic Geology (BEG). The support of this research by the Landmark Graphics Corporation via the Landmark University Grant Program is acknowledged. The publication is authorized by the director of BEG.

APPENDIX: UPSCALING ISSUES

Arithmetic Mean Versus Geometric Mean

The issue of whether the arithmetic or geometric mean should be used in vertical upscaling has been debated for more than 3 decades. The argument for using the geometric mean is that, because permeability data across entire reservoirs are log-normally distributed, the parametric statistical theory dictates that the geometric mean should be used (Aitchison and Brown, 1957). However, permeability distribution in a reservoir is not random. It is spatially controlled by depositional trends and vertically organized according to an intrareservoir-scale hierarchy ranging from sequences to a rock-fabric flow unit. Permeability data in a geologically constrained rock-fabric flow unit are different from those across the entire reservoir in mean, variance, and distribution. Petrophysical properties from wireline logs and outcrops within a rock-fabric flow unit have been shown to be nearly randomly distributed laterally and vertically (Lucia, 2007).

The difference between arithmetic and geometric means is a function of reservoir heterogeneity. A simple two-layer model (Figure 39A) was used to illustrate the difference between arithmetic and geometric means. The porosity in the upper 2 ft (0.6 m)

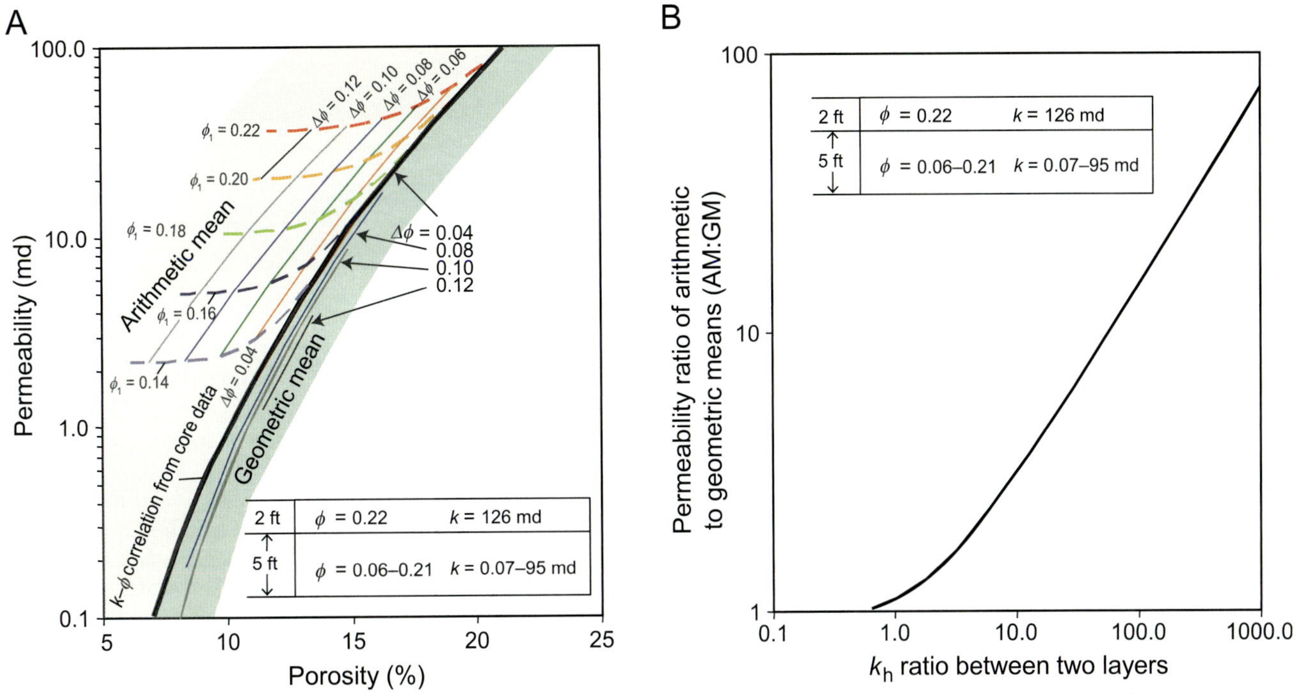

FIGURE 39. Comparison of arithmetic and geometric means in permeability and upscaling. (A) Effect of arithmetic and geometric means on upscaling. (B) Effect of permeability-thickness contrast on the ratio of arithmetic and geometric means. A simple two-layer model of porosity and permeability is used in both cases.

layer ranges from 0.14 to 0.22. For each porosity value in layer 1, the porosity in the lower 5 ft (2 m) layer was 0.01 to 0.1 less than that of layer 1. Permeability values were calculated from Lucia's correlation for grain-dominated (class 2) rocks (thick, solid black curve in Figure 39A). The arithmetic means located on the left-hand side of Lucia's porosity-permeability curve calculate higher permeabilities, whereas the geometric means on the right-hand side of the black curve (depending on the porosity difference between layers 1 and 2) calculate lower permeabilities. Note that the mean values do not follow the porosity-permeability correlation (solid black curve) and that the geometric means are closer to the correlation curve than the arithmetic means. This closeness suggests that the commonly used method of calculating permeability values from upscale porosity values using the porosity-permeability correlation is closer to upscaled permeability using the geometric mean.

The data in Figure 39A can be normalized using the ratios of means and k_h. Figure 39B shows that the ratio of the arithmetic mean to the geometric mean increases with the k_h ratio between the two layers. In highly heterogeneous reservoirs with high k_h ratios, the geometric mean can be an order of magnitude lower than the arithmetic mean.

Crossflow Effect on Upscaling

Fluid flow in reservoirs is dominated by large ordered geologic features, such as depositional trends, HFCs, stacking pattern of flow units, breccias, fractures, and others, which are predominantly layered, and the arithmetic mean is normally used for vertical upscaling according to the theory of fluid flow in porous media. Determining an average value in a block within a rock-fabric flow unit requires two upscaling steps. First, the permeability data calculated at wireline-log scale, normally vertical, are averaged at well blocks in a layer of a rock-fabric flow unit, then the block-averaged permeability data are upscaled to the rock-fabric flow unit (Figure 39). Layered (Figure 40A–C) and checkerboard (Figure 40D–F) models have been used commonly in reservoir simulation to represent two different permeability distributions and flow patterns in a flow unit. The layered model is conventional, and the checkerboard model is for near randomly distributed data. In the case of a no vertical crossflow, the water is channeled through high-permeability streaks in the layered model (Figure 40B), whereas the model becomes homogeneous in the near randomly distributed data (Figure 40E).

Note that one of the assumptions in upscaling using the geometric or arithmetic mean is a no vertical

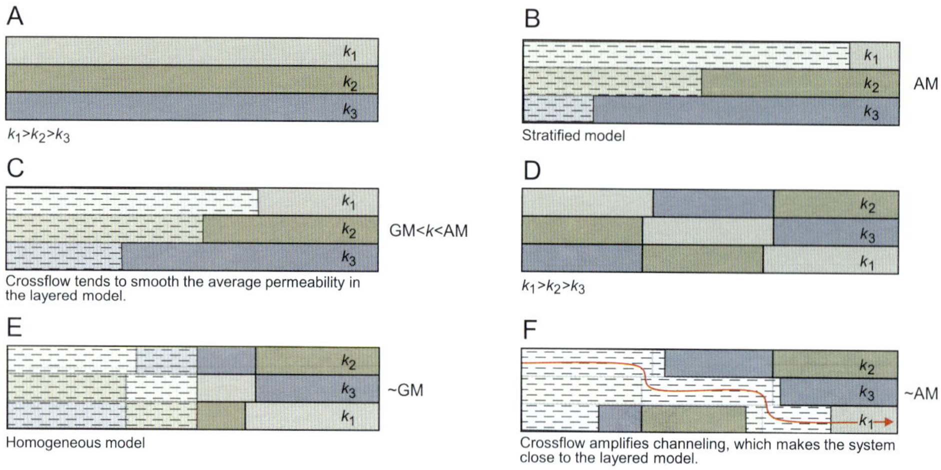

FIGURE 40. Schematic diagram showing the effect of crossflow on fluid flow in layered and checkerboard models. (A) Layered model. (B) Layered model without crossflow ($k_v = 0$). (C) Layered model with crossflow ($k_v = k_h$). (D) Checkerboard model. (E) Checkerboard model without crossflow ($k_v = 0$). (F) Checkerboard model with crossflow ($k_v = k_h$). Note that the crossflow tends to smooth the average permeability in the layered model, whereas in the checkerboard model, it amplifies channeling, making the system behave close to the layered model. AM = arithmetic mean; GM = geometric mean.

crossflow. In reality, fluid flow in a rock-fabric flow unit is dominated by crossflows because low-permeability baffles are thin and laterally discontinuous. Crossflows tend to connect high-permeability elements in a flow unit and make it behave close to the layered model (Figure 40C, F), that is, crossflows in a flow unit shift average permeability from the geometric mean to the arithmetic mean.

Upscaling from Flow Units to Flow-Simulation Model

Flow units are the building blocks for geologic and reservoir simulation models. High-frequency cycles are used to define the geologic framework for these models. In Paleozoic carbonates in west Texas, HFCs are typically upward shallowing, from mudstone at the base to grain-dominated fabric to grainstone at the top. Well-block data are mapped through a high-resolution 3-D geologic model constrained by flow units and HFCs, which are subsequently upscaled into a coarse-grid reservoir simulation model. Figure 41 illustrates how well-block data are mapped through a small part of the high-resolution 3-D geologic model (Figure 41B), which is subsequently upscaled into a column of a three-layered coarse-grid reservoir simulation model (Figure 41C). Upscaling smoothes data toward their mean and tends to shift the checkerboard model into the layered model in practice.

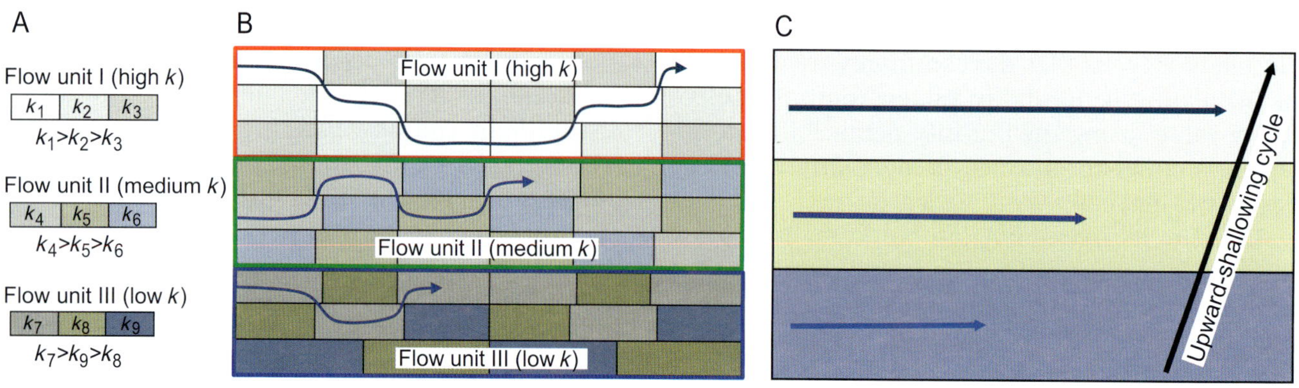

FIGURE 41. Schematic diagram showing flow-unit modeling and upscaling in an upward-shallowing high-frequency cycle. (A) Fluid flow in a high-resolution checkerboard model. (B) Fluid flow in an upscaled model.

REFERENCES CITED

Aitchison, J., and J. C. Brown, 1957, The lognormal distribution: Cambridge, England, Cambridge University Press, 1957.

Exxon Southwest/Rocky Mountain Division, 1985, Stage-I CO2 Project, Fullerton Clear Fork Unit, Andrew County, Texas: Exxon Company, U.S.A., 56 p.

Harrington, R. R., and F. J. Lucia, 2012, Integration of rock fabrics and stratigraphy for petrophysical quantification of reservoir framework, *in* S. C. Ruppel, ed., Anatomy of a giant carbonate reservoir: Fullerton Clear Fork (Lower Permian) field, Permian Basin, Texas: Studies in Geology 63, p. 49–66.

Kerans, C., F. J. Lucia, and R. K. Senger, 1994, Integrated characterization of carbonate-ramp reservoirs using Permian San Andres Formation outcrop analogs: AAPG Bulletin, v. 78, no. 2, p. 181–216.

Lucia, F. J., 1983, Petrophysical parameters estimated from visual description of carbonate rocks: A field classification of carbonate pore space: Journal of Petroleum Technology, v. 35, no. 3, p. 629–637.

Lucia, F. J., 1999, Carbonate reservoir characterization: New York, Springer-Verlag, 226 p.

Lucia, F. J., 2007, Carbonate reservoir characterization: An integrated approach, 2nd ed.: New York, Springer-Verlag, 336 p.

Lucia, F. J., and R. D. Conti, 1987, Rock fabric, permeability, and log relationships in an upward-shoaling, vuggy carbonate sequence: Austin, Texas, The University of Texas at Austin, Bureau of Economic Geology Geological Circular 87-5, 22 p.

Lucia, F. J., and J. A. Kane, 2012, Calculation of permeability and initial water saturations from wireline logs in a mature carbonate reservoir, *in* S. C. Ruppel, ed., Anatomy of a giant carbonate reservoir: Fullerton Clear Fork (Lower Permian) field, Permian Basin, Texas: Studies in Geology 63, p. 77–91.

Lucia, F. J., C. Kerans, and R. K. Senger, 1992, Defining flow units in dolomitized carbonate-ramp reservoirs: Society of Petroleum Engineers Annual Technical Conference and Exhibition, Washington, D.C., October 4–7, 1992, SPE Paper 24702, p. 399–406.

Lucia, F. J., C. Kerans, and F. P. Wang, 1995, Fluid-flow characterization of dolomitized carbonate-ramp reservoirs: San Andres Formation (Permian) of Seminole field and Algerita escarpment, Permian Basin, Texas and New Mexico, *in* E. L. Stoudt and P. M. Harris, eds., Hydrocarbon reservoir characterization: Geologic framework and flow-unit modeling: SEPM (Society for Sedimentary Geology) Short Course Notes 34, p. 129–153.

Lucia, F. J., S. C. Ruppel, R. H. Jones, J. A. Kane, F. Wang, and H. Zeng, 2006, Reservoir characterization of the Fullerton Clear Fork field, Andrews County, Texas, *in* R. M. Slatt, M. Bowman, J. Castagna, T. Good, J. Grace, R. Latimer, B. Loucks, M. Scheihing, R. Smith, eds., Reservoir characterization: Integrating technology and business practices: Proceedings of the 26th Annual Gulf Coast Section SEPM Foundation Bob F. Perkins Research Conference, Houston, Texas, December 3–6, 2006, p. 651–683.

Railroad Commission of Texas, 1986, Docket No. 8–87,428, Hearing File, May.

Ruppel, S. C., and R. R. Harrington, 2012, Facies and sequence stratigraphy: Critical tools for reservoir framework definition, Fullerton Clear Fork Reservoir, Texas, *in* S. C. Ruppel, ed., Anatomy of a giant carbonate reservoir: Fullerton Clear Fork (Lower Permian) field, Permian Basin, Texas: Studies in Geology 63, p. 5–48.

Wang, F. P., F. J. Lucia, and C. Kerans, 1998, Integrated reservoir characterization study of a carbonate-ramp reservoir: Seminole San Andres Unit, Gaines County, Texas: Society of Petroleum Engineers Reservoir Evaluation and Engineering, v. 1, no. 2, p. 105–113.

Wang, F. P., S. C. Ruppel, F. J. Lucia, R. Jones, J. Kane, and H. Zeng, 2007, Full-field modeling of the largest Lower Permian carbonate reservoir in the Permian Basin: Fullerton Clear Fork field, Andrews County, Texas: West Texas Geological Society Bulletin, v. 46, p. 4–9.